International Conference on Design & Construction of Super & Mega Yachts 2019

AA005179

Genoa, Italy
14 - 15 May 2019

ISBN: 978-1-5108-9538-6

Printed from e-media with permission by:

Curran Associates, Inc.
57 Morehouse Lane
Red Hook, NY 12571

Some format issues inherent in the e-media version may also appear in this print version.

The Royal Institution of Naval Architects
8-9 Northumberland Street
London, WC2N 5DA
United Kingdom

Phone: 020 7235 4622
Fax: 020 7259 5912

publications@rina.org.uk

Additional copies of this publication are available from:

Curran Associates, Inc.
57 Morehouse Lane
Red Hook, NY 12571 USA
Phone: 845-758-0400
Fax: 845-758-2633
Email: curran@proceedings.com
Web: www.proceedings.com

CONTENTS

Design & Construction of Super and Mega Yachts

14-15 May 2019, Genoa, Italy

PAPERS

THE ROYAL INSTITUTION OF NAVAL ARCHITECTS
8-9 Northumberland Street, London, WC2N 5DA, Telephone: +44 (0)20 7235 4622

AUTHORS' CONTACT DETAILS

Claire McKee
BCTQ
Claire.McKee@bctq.com

Sascia Sartorio
BCTQ
Sascia.Sartorio@bctq.com

Julian Smith
Cayman Islands Shipping Registry
Julian.Smith@cishipping.com

Carlo Vinelli
Isle of Man Ship Registry
Carlo.Vinelli@gov.im

Nick Danese
NDAR
ndar@ndar.com

Stéphane Dardel
NDAR
sd@ndar.com

Albert Nazarov
Albatross Marine Design
an@amdesign.co.th

Jack Wijnants
AmaSea Yachts
chairmans.office@oxbridge.eu

Robert Shaw
Shaw Yacht Design
Rob@shawyachtdesign.com

Karsten Hochkirch
DNV GL
karsten.hochkirch@dnvgl.com

Stefan Deucker
DNV GL
stefan.deucker@dnvgl.com

Oleksandr Kanifolskyi
Odessa National Maritime University
oleksandrkanifolskyi@gmail.com

Maciej Radon
DNV GL
maciej.radon@dnvgl.com

Upendra Malla
Hornblower Cruises
au.upendra@gmail.com

Matteo Magherini
Lateral
mmagherini@lateral.engineering

Alex Meredith
Lateral
amhardy@lateral.engineering

Adrien Thoumazeau
Lateral
athoumazeau@lateral.engineering

Andreas Witschel
Schottel
SThewalt@schottel.de

Muhammad Taha Ali
DAMEN Shipyards Galati
tahaali892@hotmail.com

Ziya Saydam
Piri Reis University
zsaydam@hidro-teknik.net

Mincui Liang
Newcastle University
M.Liang3@newcastle.ac.uk

Olaf Doerk
DNV GL
olaf.doerk@dnvgl.com

Steve Beech
Safeguard Helideck Certification
steve.beech@safeguardengineering.co.uk

Mustafa Insel
Hidroteknik Nautical Design Technologies
minsel@hidro-teknik.net

Turhan Soyaslan
Soyaslan Design
turhan@soyaslanmarine.com

Mark Downer
Griffon Hoverwork
mark.downer@griffonhoverwork.com

Stefan Semrau
DNV GL
stefan.semrau@dnvgl.com

NEW BUILDS AND THE IMPORTANCE OF THE SPECIFICATION

C N McKee, S F Sartorio, BCTQ, UK

SUMMARY

The tender and contract technical specifications are discussed and their importance to the overall success of a new build yacht project explained. We explore the way in which the technical specifications and technical advisors fit into the overall project management processes of a new build, as well as the technical specifications interconnected relationship with other important documentation. We discuss some examples where the technical specifications made a large difference to the overall success of a project and how simple details can have significant knock on effects. For both the tender technical specification and the contract technical specification, achieving an appropriate level of non-subjective detail is key and we show that, whilst projects have different overall processes, the technical specifications lie at the heart of successful and unsuccessful ones.

1. INTRODUCTION

The technical specifications are the most important documents during a yacht build as they can dictate the overall success of a project. Success, in this case, measured by the balance of cost, quality and time to achieve the owner's expectations. This includes minimising change orders and permissible delays whilst balancing the overall cost of the project.

The way in which the client's statement of requirements is analysed by the owner's representative and technical advisors, then translated into the design brief and technical specification, is key. However, there is more to the process than the design brief and technical specification alone as they must tie into a number of other important documents and secondary specifications.

In this paper, we explain the significance of the two most important technical specifications: the tender technical specification and the contractual technical specification. We also explore the way in which they fit into the overall project management process as well as their interconnected relationship with other important documentation.

For both the tender and contract technical specifications, achieving an appropriate level of non-subjective detail is key: too vague and it can be open to the wrong interpretation, too specific it can be overly restrictive. We therefore discuss a number of examples where these specifications made a substantial difference to the overall success of a project and how simple details can have vast knock on effects.

We show that, while projects can have different overall processes, the technical specifications are key to their overall success. Therefore, to achieve robust specifications, contracts and understanding one must involve technical advisors as early as possible in the process.

We look at how the technical specifications can affect a project in one of three ways: performance, design or process and we show how these different details of the technical specifications weave their way through the whole build process of a yacht.

2. CLIENT STATEMENT OF REQUIREMENTS

All yacht new build projects start with the client's statement of requirements. This process can vary significantly from project to project. Each client will have different previous yachting experience and expectations. Some clients will just have a sense of the price or size of vessel that they want whilst others will focus on aesthetics or specific features such as the tenders and toys that they would like to carry on-board. However, with good guidance from the appointed technical advisors their expectations in terms of performance, technical details and design can be understood and translated into an effective design brief.

3. TENDER TECHNICAL SPECIFICATION

Within a project there are two phases of technical specification, the first being the tender technical specification, followed later by the contract technical specification. Overall the objective of the tender technical specification is to capture the client's statement of requirements and freeze the design at its feasibility stage.

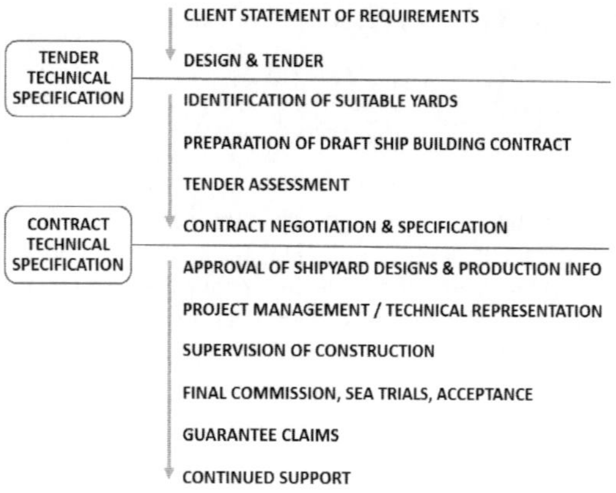

Figure 1: Overall Procurement Process

The level of detail in a tender technical specification can vary considerably but it should generally:

- Freeze design at feasibility stage
- Capture the client statement of requirements
- Define features
- Define operational requirements
- Define regulatory requirements
- Establish the basis of cost and programme requirements
- Establish the basis of the project risk analysis
- Mitigate potential basic design flaws

Whilst the main objective for the tender technical specification is to summarise the concept design features of the future vessel, it is used as part of the competitive tendering process utilising the client's statement of requirements to identify suitable yacht builders.

It will also intrinsically influence the construction phase and lifecycle of the vessel. Indicative basic parameters, such as length, breadth, depth, draught and installed power are decided and key design features will be identified at this stage that can only be partially modified, in the next phase of the design, without a major impact on costs and timescale.

For example, as naval architects know well, the basic hull length to breadth ratio has massive implications on speed performance and stability. But also something as simple as a concept external design with a large amount of glazing will have a significant impact on cost as well as structural and fire safety implications. A preliminary review of the concept external and internal design sketches, if available at this stage, is beneficial as the owner's advisors may already start translating those concepts into sensible technical features.

One of the main risk areas that the tender technical specification along with the technical advisor should help mitigate are a lack of consolidated and unrealistic client's statement of requirements, which can lead to unrealistic time frames, expectations and unexpected costs.

As all projects vary in terms of the involvement of technical advisors, it is important that at this stage it is ensured that the client's statement of requirements, concept design and the design brief are technically feasible. If they were developed without a technical advisor involved, they may be not only be technically unfeasible but also contradictory.

As a result, the challenge is balancing a very complex set of variables. Sometimes the most basic ones set by the owner can be conflicting, such as the speed and length or cost and size. Once the initial particulars have been decided, the more detailed requirements need to be determined. It does happen with yacht projects that through the initial phases and concept design, the main focus can be on the aesthetics and owner's areas of the vessel. At times the technical feasibility, regulatory constraints as well as operational needs of the vessel are simply ignored. This is often down to a lack of technical advisor involvement and instead the project being driven by interior and exterior designers.

The operational requirements needed by a yacht may not be something that the owner thinks of. Returning to our earlier example of a request for large glazing, there will also be operational implications such as access and available fresh water for cleaning. At this point a reality check may be required. The owner will need to understand the implications of certain features they have requested. For example, the owner's expectations of tenders / watercrafts to be carried on-board could have major impacts on the general arrangement plan (GAP) and aesthetics but also other technical aspects such as sufficient fuel provision, maintenance requirements, launching and recovering methods. The earlier these things can be addressed in terms of feasibility or making provisions for them, the better.

It is the technical advisor's responsibility to make sure that the client's statement of requirements and design brief are translated into a tender technical specification that is technically feasible. Whilst all aspects do not need to be designed in detail at this point, it does need to be ensured that the vessel being tendered for can be built, operated and classed and flagged to the owner's desired flag and class notation.

4. IMPORTANCE OF FLAG AND CLASS

Flag and classification of a vessel should be selected before the tender specification stage. Alternatively, a number of different options may be investigated and indicated by the owner's advisors. This exercise should come as part of the client's statement of requirements and design brief but again depending on the involvement of a technical advisor, this may not have been properly addressed.

Early engagement with the flag administration and the classification society is of paramount importance. Their view and requirements can have a significant impact on the design of the vessel. The location and size of stairwells as well as the locations of bulkheads and fire boundaries will have a significant impact on the general arrangement plan as well as the structural arrangement and may affect the feasibility of a particular concept design.

Clients can be led to believe that if a vessel is flagged and classed it will be of an acceptable standard and therefore a detailed specification will not be required. But whilst appropriately flagging and classing a vessel is very important it does not replace a detailed specification. Classification and flag requirements will have very little to do with the finished quality of a yacht. They will only ensure that it is built to a certain structural, stability and fire protection standard but it will not dictate the quality of finish or functionality of the future yacht. Whilst classification societies have additional notations, such as ABS COMF(Y) + [1] and LR PCAC [2] they need to be specifically referred to in the specification and often added to.

In contrast, some yacht owners will say that they intend to only use their yacht privately and therefore ask for more relaxed flag requirements and potentially not even class their vessel. It is important to make clear very early on to the yacht owner that this can have a serious impact on the value and possibly the safety of the vessel. We will often recommend that an owner builds their yacht to an appropriate commercial standard and then operate it privately as they wish. If a specification has been written without the intention to operate a vessel commercially, it will be very difficult if the client changes their mind at a later date.

Specifications not correctly identifying an experienced class and leaving too much freedom to the yacht builder can be risky. Experienced class and flag can help in solving problems and foreseeing potential issues. Risk Based Design (RBD) techniques may be used for designs deviating from specific rules and regulations as well as for complex and novel designs.

For instance, it happened that an inexperienced builder chose a classification society for reasons of cost and familiarity without properly considering whether they had enough expertise with a new type of propulsion. In the post-contract stage class required an RBD approach which led to an increase of engineering design cost and major delays. A proper risk analysis should have captured that an inexperienced builder should not have been allowed to choose an inexperienced classification society in the first place.

This example shows how the contract technical specification can impact the process rather than the design. In particular, if the tender technical specification could be kept generic, the contract technical specification should have captured the requirement of opting for an experienced classification society.

Once the selection of a classification society and flag administration has been completed, it is then important to decide the operational area, passenger and total number of persons on board. For Red Ensign Group flagged vessels this will dictate whether a vessel falls under REG Yacht Code Part A (formerly Large Yacht code or LY4) or REG Yacht Code Part B (formerly Passenger Yacht Code or PYC) [3].

REG Code Part A, Large Yacht Code, is for vessels with 12 passengers or less, whilst REG Code Part B, Passenger Yacht Code is for up to 36 passengers. The different implications for the design and the specification, if they fall under the REG Code Part A or B, will be substantial. The additional requirements for the REG Code Part B include much more onerous fire protection, including non-flame spread materials, much more onerous stability and/or lifesaving appliance requirements which can have a significant impact on the layout and cost of the vessel. The operational area and total number of person on-board can further impact requirements for life rafts versus life boats and enhance stability measures.

A yacht builder can apply to flag and class for a number of exemptions or operational limits. Therefore, it could be important to stipulate any allowances or procedures for obtaining the owner's permission prior to this occurring.

5. CONTRACT TECHNICAL SPECIFICATION

The contract technical specification will be used once the yard has been selected. It forms an integral part of the contractual documentation package and will be the cardinal point of the overall contract negotiations. As a consequence, it must be congruent with and properly linked to the other documentation, in order to ensure there are no conflicting and confusing statements or requirements.

The contract technical specification will build on the tender technical specification. Therefore, it will:

- Capture the client statement of requirements
- Define features
- Define operational requirements
- Define regulatory requirements

In addition, it will also:

- Freeze the design
- Define in detail the technical deliverables
- Define the standard of workmanship

- Define where responsibilities lie
- Dictate the approval of yacht builder designs and production information
- Determine the project management and technical representation and who is responsible for the supervision of the construction.
- Include details of owner supplied equipment
- Determine the ability to inspect and attend tests as well as the final commissioning, sea trials and acceptance.

By experience, the level of detail in a contract technical specification can vary considerably. That of a <50 m private yacht built by newcomers in remote places is inevitably less detailed of that for a PYC +100 m built by an experienced yard in Europe. However, both owners expect their yacht to have the highest possible quality, according to the statement of requirements and to get it on time and on budget. Thus, the principles behind the specification should not be affected by the project complexity or the yacht builder's capabilities.

The shape and form of the contract technical specification can also vary significantly. Either a yacht builder or the owner's technical team will have their own standard and layout. BCTQ have a custom contract technical specification developed and refined over many years. Depending on the project, this can act as the basis for the contract technical specification and the builder can then adjust it to fit with their practices. This could be the case with less experienced builders. Many yacht builders will have their own layout based on the preferred work breakdown structure (WBS) as it can often be the case that the production process and cost structure will be tied closely to the technical specification. For example, each system can be linked to a specification chapter and each chapter or section item could have a separate commissioning form.

A worldwide used 3-digit decimal classification system for the maritime and offshore industry is the SFI Group System. This has a WBS consisting of the following primary groups [4]:

- General
- Hull Systems
- Cargo Equipment
- Ship Equipment
- Crew and Passenger Equipment
- Machinery Main Components
- Systems for Machinery Main Components
- Common Systems

It is common to have specifications that follow similar sections based on what is applicable to the type of vessel.

Whatever form the contract technical specification takes, it is very important as it will form the back bone of the build process and dictate the quality of the final product. Even seemingly small details can have a significant impact.

For example, on a 105 m yacht project, the ceiling free heights indicated in the specification turned out unfeasible due to not enough space being left for services in the initial design. The investigation to increase ceiling heights meant that the overall height of the vessel needed to be increased, this meant an increase in breadth which in turn meant an increase in power requirements. These knock on effects meant a significant increase in cost and delays.

Another seemingly small detail, is the colour of the hull. In another large yacht project, the client initially opted for a black hull. This had a significant impact on the HVAC compared to the reference project, due to the thermal absorption. Space for the HVAC system is often critical and any increases could significantly impact the GAP. Therefore, even something which appears to be a fine detail such as the colour of the hull needs to be considered and carefully risk assessed at the specification stage of the project. Depending on the design aspects, this process may lead to increased design margins, specific in-depth analysis or preliminary tests.

Therefore, to get an accurate overall build cost and schedule, it is critical that every detail in the contract technical specification is correct and, where generic indications are provided, all the technical implications have been carefully considered by the owner's technical and legal advisors in collaboration with the yacht builder as applicable. This can be through experience, detailed risk assessment or initial testing and analysis. Spending the time and money up front makes significant time and cost savings on an overall project.

It is imperative that if there are specific requirements these should be noted in the specification. It may sound obvious but it is often overlooked where it is assumed common sense or good practice by the detailed design team and yacht builder will prevail. However, sense and good practice are subjective and may be influenced by different priorities, familiarity, engineering versus aesthetics or simple cost versus quality. Therefore, where specific requirements exist, the specification should include sufficient details to ensure that subjective interpretation is not possible or that adequate means of minimising risk are in place.

Aesthetics is an area where subjective views are apparent. If the design waterline is important from an aesthetic perspective this should be highlighted. For example, with a 100 m yacht it was unclear in the specification if the bulbous bow should be submerged over a range of draughts and trims. The yard designed the vessel such that the bulb would only be submerged at the maximum design draught. This meant that the bulb was rarely submerged at operational draughts, therefore the yacht owner was left sailing around the world with an inefficient but also unaesthetic fouled bulb. If the design waterline is important from an aesthetic perspective then this should be stated.

On the subject of waterlines, how the position of boarding platforms relates to the design waterline is essential: get this wrong and you may end up with a too great a step up from a tender or a yacht owner getting wet feet. Stating exact expectations is therefore crucial.

6. IT'S ALL IN THE DETAILS

The most important aspect of any successful specification is understanding the potential impact of specific details on the overall project development. The objective is to find the best way to express detail and understand its impact.

The specific way in which the technical specification details drive a project can take one of three forms:

- Performance Based
- Design Based
- Process Based

If a specification detail is performance based, it will describe a specific desired outcome, such as speed, range or noise and vibration limits and these will often be linked to the build contract and performance penalties such as liquidated damages.

If a specification detail is design based, it describes what the end product will look like or should be (in the case of a specific requirement). This could include reference to the general arrangement plan and interior design drawings. It may also reference secondary specifications such as interior outfit specifications. In yacht projects, the design details laid out in the specification can be the most important aspect in achieving the client's expectations of the final product.

Finally, if a specification detail is process based, it describes a process to drive an end result. This is where reference to yacht builder standards, ISO and other standards, paint specification and specific testing requirements come into play.

Some details may have multiple implications, due to their intrinsic complexity or unforeseeable knock on effects. These details should be properly risk assessed at the beginning and carefully monitored during the project development.

Once the content of the specification has been identified, achieving an appropriate level of detail is key: too vague and it can be open to the wrong interpretation, too specific it can be overly restrictive.

For example, when describing a pump for a specific requirement, one could either:

- Say *'an adequate pump'*
- Detail the function of the pump such as its capacity
- Give a specific model and make of a pump

All three options have their risks. Saying *'adequate'* leaves the situation left open to interpretation: what is *'adequate'*? But to give the specific capacity of the pump, what if the requirement changes and that capacity is no longer sufficient? Stating a make and model: what if that make and model is discontinued or turns out to not be appropriate and other aspects of the system have been designed to that specific model?

Therefore, it is important for a detail to fall within the appropriate category. For example, for a swimming pool pump, it could be appropriate to have it performance based and link the pump to the desired performance: *'It must be able to fill the pool in one hour'*. This makes it is clear what is *'adequate'* and if the volume of water changes, the pump can be specified accordingly.

In another example, where one is specifying the petrol filling pump for the tenders and jet skis, it could be appropriate to have it design based and state a specific make and model as it is a relatively standalone and unchanging function. The owner's team may have experience with that specific pump and want to use it again. This would simplify the procurement process and fix the price making both the yacht builder and the owner happy.

However, the surface finish of a yacht may require both design and process based details. For example, one can provide specific design details for the finish such as faring targets over a flexible batten length or long and short wave surface texture targets for the paint finish. For the surface preparation of the steel underneath the paint and filler, processed based details are required such as ISO standards for steel preparation.

Details are not purely technical, the format of the specification also plays an important role. Within the industry it is understandably common to refer to an existing specification for a new build. It is therefore a common occurrence to have copy and paste mistakes. Such mistakes are usually captured during the review process but may potentially lead to gross embarrassing errors and non-disclosure agreement (NDA) breaches.

Good practice is to avoid repeating requirements in different sections of the specification and instead relying on proper cross-references within the document. This will prevent consistency mistakes, optimise the dimension of the document and univocally define any potential reference cited in other associated documentation.

In general, applying standard quality assurance (QA) procedures will prevent a number of inconvenient mistakes. The importance of legal advisors is often not stressed enough. Following a proper cost-risk analysis, legal advice is extremely important and is more effective the earlier they are involved in a project and the more efficient when clear lines of communication between them and the technical advisors are established.

7. ASSOCIATED DOCUMENTATION

The contract technical specification forms an important part of the negotiation process and as a result, it links with a number of important documents. These include but not limited to the build contract, contract drawings, protocol and acceptance forms, makers list, design and build schedules, inspection and testing plan, stage payment profile and variations form.

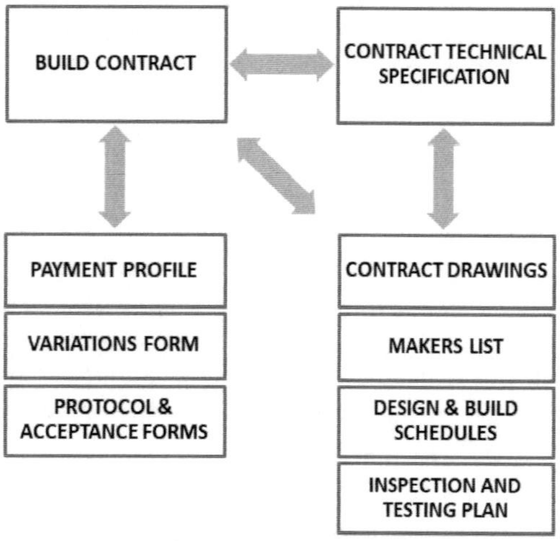

Figure 2: Associated Documentation Relationship

The link and consistency with these documents are of fundamental importance. The level of detail for specific disciplines or single topics should be consistent with the contractual priorities and design constraints.

The contract technical specification will form a complex interlacing with the documentation package associated with the build of the yacht. Therefore, it must be congruent with the other documentation in order to ensure there are no conflicting or confusing statements and requirements.

In terms of the overall success of a project, the link with the build contract could be one of the most important roles of the contract technical specification. The vessel performance criteria set out in the contract technical specification will be linked to non-performance liquidated damages through the build contract. This would include the liabilities if anything from speed, range, deadweight, draught to noise and vibration targets are not met. This linking is imperative because a contract technical specification is only as good as the consequences of noncompliance.

Despite the good initial intentions usually expressed by both parties, the project process is often affected by pressing deadlines and budget restrictions. The specified liabilities are the last resort but possibly the most powerful instrument the yacht owner's team have to assert their instructions. Often the liabilities also become the driver for key decisions with a direct or indirect impact on performance and design based details as well as process based details (general quality of the build, yard personnel health and safety, yard environmental impact).

If it is clear that specific noise and vibration targets will be hard to meet, the yacht builder could take the negotiation of the target values as a calculated risk based on the cost of not complying. If certain targets are critical to an owner, it is important that liquidated damages or vessel rejection criteria reflect this as even if it is included in the contract technical specification, it does not mean it will definitely be met in the complete vessel. In fact, it could be cheaper for the yard to pay the liquidated damages than it is for them to meet the targets.

Equally the opposite could be true, that if a yacht owner thinks that something is a nice to have but it is unknown if the yacht builder can meet the targets or achieve the expected results. If the liquidated damages are very high, then the yacht builder could add a considerable extra cost onto the vessel to mitigate the risk of non-compliance. Therefore, it is important to understand what is most important to the owner.

For example, if there is a beach area where proximity to machinery spaces means that there is a risk that noise and vibration levels could be higher than other passenger areas, it may make sense to relax the limits if the yacht owner may not require that space to be quiet. Especially if they intend to be playing loud music and have the beach doors open the majority of the time.

It is also important for the testing procedures to match with the operational use of the vessel. If, for example, the owner has noise level expectations when the beach doors are open, it is important for testing limit procedures to include realistic operational situations. If noise limits are met with the doors closed but the owner expected certain levels with the doors open, those limits are irrelevant. Of course, getting the yacht builder to agree to open door noise tests and creating a controlled environment is another matter.

The contract technical specification will also link with the build contract through the contract design drawings. These can be a very important part of the documentation but it is vital to be clear what is included as part of the package. As renders have been getting more and more detailed in recent years, it can be misleading what will be included with the final yacht, or what the final yacht will look like. Even the smallest of details can cause problems.

It has happened that an owner has seen an exterior render of a vessel which included specific outdoor furniture. A designer had included a certain table as part of the overall exterior look of the vessel. The owner particularly liked that table and assumed that it would be included in the yacht. What is included needs to be clear and if the owner wants a specific item it needs to be included in the provisional cost (PC) sums by the owner's team.

Another example where renders can be misleading, particularly exterior renders, is the absence of safety railings. Owners can often be quite disappointed when the slick profile of the vessel is spoiled with a number of unaesthetic railings. It is important to make it clear to the client what is indicative of the look of the vessel through a render and what is going to be achieved and received in reality. If it is clear that the look of the safety railings (or their absence) is of utmost importance to the client then details need to be indicated in the specification to achieve this. This can include: minimised profiles, glass railings and removable railings (where appropriate). This will all likely come at an extra cost. It is important to understand what the client is expecting and willing to pay for at the specification stage and then stipulate this as part of the price and expected outcome.

Navigation masts, life-rafts, lifeboats, cranes, mooring equipment and a helipad have a significant impact on the external design. Last but not least the decks height distribution, a crucial detail that exterior designers, especially if inexperienced, may neglect, making yachts look more streamline than is possible to achieve. This may also conflict with the owner's expectation of ceiling heights inside the yacht.

A timely pre-engineering of the design is always recommended and the significant impact that it may have on both the internal and external design must always be considered.

8. CHANGE DOES OCCUR

Despite the best efforts of the yacht owner's team and advisors, change can and does occur after the specification and build contracts have been signed.

It is important to remember that during the specification stage, there can be immense time pressure for a contract to be signed for a number of reasons. The owner being keen to see their project progress, a build slot needing to be reserved or a regulatory deadline could be looming putting pressure on a keel laying date.

The time for the development of a design must be enough to develop and draw up a solid contract technical specification, which requires key bits of the design to be fixed. It is also absolutely crucial that there is allowance for some change. For example, system drawings that are signed along with the contract technical specification do tend to change following a consultants plan appraisal or during detailed design. Therefore, a good technical specification needs to allow for change and must be compatible with a procedure for change which includes the variations form.

A variation form should ask what is changing and why. It needs to assess the time, cost and technical impacts of the change. It should address implications of other contractual obligations such as weight targets. The yacht builder needs to state all the impacts of a change or else they will still be held to the original contractual obligations and targets.

If throughout a project the owner asks for changes and the yacht builder does them without a variation form and eventually targets or deadlines are not met, the yacht builder can try and use those changes as an excuse for not meeting such targets. Consequently, it is in everyone's interest to follow correct procedure for all changes.

The contract and specification should allow the owner to have sufficient influence on the developing vessel, including time and priority to their technical advisors, to make sure that the interpretation of the contract and any changes to the build / design, as an evolving entity, is to the owner's advantage and not simply allowing the yacht builder to rewrite the specification.

9. CONCLUSIONS

The key tools that a yacht owner can use to mitigate risks within projects are the tender and contract technical specification and their link with the build contract and other documentation. The joint effort of the owner's team, technical advisors, legal advisors and yacht builder become the most critical and the details within these documents become the key drivers for success.

The tender technical specification ensures technical feasibility and makes sure that project risks are kept to a minimum. It establishes the basis of the project risk analysis and intrinsically mitigates potential basic design flaws. Whilst the main objective for the tender technical specification is to summarise the concept design features of the future yacht, it is used as a basis for yacht builders to tender for the build and will impact, whether it be directly or indirectly, on the construction phase and lifecycle of the vessel.

The contract technical specification forms the back bone of the build process. It dictates the quality of the final product and forms a key part of the negotiation process through its links with a number of important documents. Therefore, to get an accurate overall build cost and schedule it is critical that every detail in the contract technical specification is correct and, where generic indications are provided, all the technical implications have been carefully considered whether they are design, performance, processed based or any possible combinations. Whether this is through experience, detailed risk assessment or initial testing and analysis, spending the time and money up front makes significant time and cost savings on an overall project.

The landscape of yacht building is continuing to change and the ways in which yachts are being built continues to evolve. Yacht owners' demand for high-tech on board is growing. Semi-custom yacht production, once reserved for 'smaller' yachts, is now growing larger and larger. In addition, the upper limits of the largest yachts are also increasing, so the complexity of vessels is broadening and along with it, their risks.

This means the complexity of the specifications grows too and the job of the technical and legal advisors becomes even more critical to ensure the overall success of a yacht project.

10. REFERENCES

1. American Bureau of Shipping, 'Guide for Passenger Comfort on Ships', ABS, September 2014 (Updated January 2015).

2. Lloyd's Register Rules and Regulations for the Classification of Ships, July 2017.

3. Marine and Coastguard Agency, 'Red Ensign Group Yacht Code', MCA, January 2019.

4. The Ship Research Institute of Norway, 'SFI Group System', 1973.

11. AUTHORS BIOGRAPHY

Claire N. McKee holds the current position of Senior Consultant Naval Architect at BCTQ. She is responsible for a wide variety of consultancy project including the design development and project management of a number of large yacht projects between 100 and 140 metres. She has a range of expertise including international rules and regulations, concept design, stability analysis and plan approval. In the last few years Claire has been on the PYC industry working group.

Sascia F. Sartorio holds the current position of Technical Manager and Principal Consultant Naval Architect at BCTQ – Southampton. Along with day-to-day office management duties, he is responsible for procurement technical advice, concept and detail design, surveys as well as plan approval on behalf of the owner. His previous experience includes basic and detailed design in a variety of disciplines across different sectors of the marine industry, including large displacement and semi-displacement yachts for Benetti, CRN, Fincantieri, ISA and Perini Navi.

BUILDING TO THE REG YACHT CODES

J C Smith, Cayman Islands Shipping Registry, UK
C Vinelli, Isle of Man Ship Registry, Isle of Man

SUMMARY

The Red Ensign Group (REG) has recently completed the development of an updated regulatory framework for Yachts, entitled the 'REG Yacht Code', which entered into force on the 1st of January 2019.

The Code's aim is to further develop the well-established industry standards of the Large Commercial Yacht Code (LY3), combining the lessons learned from almost 20 years of regulating the Large Yacht sector since the first version of the Code was published in 1997 & Passenger Yacht Code (PYC) since the 1st edition was published in 2010. The revised text makes larger use of industry best practice, international standards and the IMO's overarching remit for the provision of increased 'Goal Based Standards' with the intention of allowing room for more flexibility and innovation in the design and construction of Large Yachts.

Here, the principle changes to the Code are presented in addition to the benefits and challenges that are associated with achieving compliance. Finally, the key technical considerations during the design and construction process when building to Parts A and B are discussed.

NOMENCLATURE

AD&A	Alternative Design & Arrangements
DLLR's	Davit Launched Liferafts
FTP	Fire Test Procedures Code 2010
IS Code	Intact Stability Code, 2008
LY3	Large Commercial Yacht Code
MLC	Maritime Labour Convention 2006
NFPA	National Fire Protection Agency
PY-1	Passenger Yacht restricted to Prescribed International Voyages
PY-2	Passenger Yacht restricted to operate no further than 60nm from a Safe Haven and 20nm from land in favourable weather
PYC	Passenger Yacht Code
PY-P	Pleasure Vessel not Engaged in Trade
PY-U	Passenger Yacht Unrestricted
REG YC	Red Ensign Group Yacht Code
SOLAS	International Convention for the Safety of Life at Sea, 1974

1. INTRODUCTION

Part A of the Code is applicable to Yachts which are 24 metres and over in load line length and which do not carry cargo and do not carry more than 12 passengers or if built before 21 July 1968, 150 gross tons and over according to the tonnage measurements applying at that date. Those elements of the Code which are considered to constitute an equivalent to SOLAS are based on Cargo Ship requirements.

Part B of the Code is applicable to Yachts of any size carry more than 12 but not more than 36 passengers on international voyages. Those elements of the Code which are considered to constitute an equivalent to SOLAS are based on the requirements for Passenger Ships carrying no more than 36 Passengers.

Both Codes contain provisions for Motor and Sailing Yachts.

2. CODE DEVELOPMENT

Recognising that in some areas, the Large and Passenger Yacht Codes required amending in order reflect regulatory changes to the underlying conventions and accommodate developments in design trends and technology, the Red Ensign Group in close collaboration and consultation with the Large Yacht Industry undertook to perform a comprehensive review and update of the existing framework.

The process involved the formation of four Industry Working Groups consisting of REG Members, Classification Societies, Aviation Inspection Bodies, Designers, Builders and Management Companies with separate groups being formed to address the Large and Passenger Yacht Codes, the carriage of more than 120 Persons onboard Passenger Yachts and Helicopter Landing Areas. The deliberations and

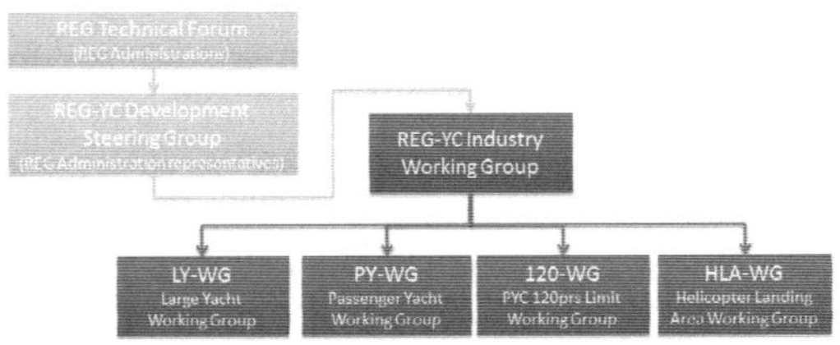

Figure 2.1 – Overview of Code Development Process

The Government of the United Kingdom has agreed in accordance with Article 8 of the International Convention on Load Lines, 1966, regulation 1/5 of the international Convention of the Safety of Life at Sea and Article 9 of the International Convention of the Standards of Training Certification and Watchkeeping for Seafarers 1978 that the Code constitutes an equivalent arrangement and has notified the IMO Secretary General accordingly. The acceptance of these equivalent arrangements is extended to the United Kingdom Overseas Territories and the United Kingdom Crown Dependencies [1].

The IMO are now moving to four yearly amendment cycle. The intention is for the new Code to follow this cycle, although a series of Corrigenda will be published at appropriate times based on feedback and experience gained during the early stages of its implementation.

3. STATUTORY LIMITATIONS REGARDING EXEMPTIONS AND EQUIVELANT ARRANGEMENTS

The extent to which any Flag Administration is able to grant an exemption to a Statutory Rule requirement is limited to the provisions which are included in the underlying International Conventions. Where it has been identified that design details are not in compliance with the applicable requirements of the YC and an exemption could not be granted due to the absence of any suitable provisions, the Administration should be approached with a view to investigating the feasibility of accepting departures from prescriptive rule requirements based on achieving an equivalent level of safety.

Whilst a reference to proposals regarding the acceptance of alternative standards, which may be considered to achieve an equivalent level of safety were previously included in both the Large and Passenger Yacht Codes, Parts A and B of the REG Yacht Code now provide additional clarification regarding the minimum expected criteria to ensure that the balance of safety is properly addressed. The Codes now clarify that such arrangements are to be based on the underlying IMO Conventions and ensure that additional mitigating measures are provided where a departure from prescriptive requirements is intended. Additional equivalence considerations are subsequently provided in the applicable Chapters. If necessary, the acceptance of such arrangements is subject to the appropriate Engineering Analysis and testing to the satisfaction of the Administration. Copies of the final documentation confirming acceptance by the Flag State should be kept on board for the life of the vessel.

It is important to note that the acceptance of an equivalent arrangement by one Administration does not provide a guarantee that the same arrangement would be automatically accepted by another Flag or Authority, hence such arrangements may have the potential to limit opportunities in relation to Change of Flag and can at times present issues in relation to Port State Control Inspections.

4. PRINCIPLE REGULATORY CHANGES

4.1 THE USE OF ALTERNATIVE DESIGN AND ARRANGEMENTS (AD&A's)

In cases where a departure from prescriptive requirements is considered to be significant, the Administration may conclude that a more in-depth study of a particular proposal is appropriate. It is important to note that in the case of an Alternative Design and / or Arrangement, the scope of work to be undertaken both by the Designers / Builder and the approval authority (i.e. the Administration or Classification Society) is substantially in excess of that which may be required to support the acceptance of what was considered to be less significant and does not automatically result in a particular proposal being accepted. The responsibility to determine at exactly which point an alternative / equivalent arrangement is significant enough to warrant the depth of study required for an AD&A ultimately rests with the Administration. Consequently, where a departure from Regulatory requirements is identified during the design and / or production process, it is essential to engage with the Administration at the earliest possible opportunity in order to agree the technical details to be covered during the study and discuss the feasibility of the proposal being accepted upon satisfactory completion of same. It must however be appreciated that the final answer regarding acceptance or otherwise could not provided until such times as the study is complete and has been reviewed in detail. The commercial risks associated with this approach should therefore be recognised and understood early in the design development phase.

The methodology for AD&A's is limited by the extent to which the International Conventions allow. Only those chapters of the Codes, which contain provisions for AD&A's may apply such principles. In order to further support the process, the IMO Guidelines on Alternative Design and Arrangements for SOLAS chapters II-1 (Structure, Sub-division, Stability, Machinery and Electrical Installations) and III (Life-Saving Appliances and Arrangements) (MSC.1/Circ.1212) and Fire Safety (MSC/Circ.1002) are referenced in the YC and should be consulted during the design development phase.

Upon the satisfactory completion of the approval process, the Administration will confirm its acceptance of any such arrangement to the IMO via a formal notification process, who will in turn circulate the details of any such acceptance to all contracting Governments. A copy of the documentation, as approved by the Administration, indicating that the AD&A has been accepted shall also be carried onboard the Yacht.

Finally, in the event that any of the assumptions and operational restrictions that were stipulated in an AD&A are modified for any reason, the engineering analysis is to be suitably amended in order to reflect such modifications and be re-approved by the Administration.

4.2 SWIMMING POOLS AND JACUZZIS – EFFECT ON STABILITY

The previous requirement for Swimming Pools and Jacuzzis to be drained within 3 minutes when fully charged with water has been retained in the REG YC. However, the requirements in LY3 to consider the mass of water and its corresponding free surface effects in both the intact and damaged conditions is applicable to Yachts certificated in accordance with the REG YC irrespective of whether such a drainage capability has been provided. Where it is determined that a Swimming Pool and / or Jacuzzi needs to be emptied in order to comply with a certain intact and / or damaged stability condition, a separate section is to be included in the Stability Information Booklet clearly documenting this matter with the following note added: "In this Loading Condition, the vessel may not have its (Swimming Pool or Jacuzzi) full, due to insufficient stability"

4.3 REQUIREMENTS FOR THE PART B VESSELS REGARDING THE CARRIAGE OF MORE THAN 120 PERSONS

In response to the ever-increasing size of Mega Yachts and Industry feedback concerning service level expectations, a Working Group was formed to consider the additional risks associated with increasing the existing 120 Person limit for Passenger Yachts. It is important to note that this initiative was undertaken due to a demand for the increase in Crew rather than Passenger numbers. The group undertook to conduct a gap analysis between the 6th Edition of the PYC and the corresponding requirements of SOLAS as applicable to Passenger Ships carrying no more than 36 Passengers. (SOLAS <36), the Loadline and STCW Conventions in order to identify any significant variations in regulatory requirements. One such variation was subsequently identified as SOLAS II-1/8.1 & II-2/21 and the requirements for Safe Return to Port (SRtP). Recognising that compliance with these requirements places a significant technical demand on the design, construction and operation of such vessels and that Passenger Yachts may frequently operate much closer to shore than commercial Passenger Ships, a series of alternative risk mitigating measures were developed with the balance of safety being addressed via an operational restriction, which is aligned with the Short Range Yacht concept available for Vessels certificated in accordance with Part A of the Code.

For vessels operating no more than 60 nautical miles from a safe haven, these requirements (in addition to the carriage of a Doctor and Lifeboats) subsequently include compliance with the Maximum Floodable Length criteria in Part VI and Enhanced Survivability criteria contained in Part VII. Whilst only applicable in SOLAS to Passenger Ships carrying more than 36 Passengers, Part B also requires additional protection measures for internal and external escape routes and assembly stations via the provision of Low Location Lighting in crew areas and fire rated glazing for those windows facing external escape routes. Additional requirements for Lifebuoys, Immersions Suits and the provision of a Flooding Detection System have also been introduced.

Noting the importance of providing an option for Passenger Yachts carrying more than 120 Persons to operate without restriction, the SRtP requirements were introduced for those vessels, which are operating in a commercial capacity. PY-P Yachts operating in a private capacity have no regulatory need to comply.

4.4 COMMON ANNEXES

Recognising that a number of safety critical features, which were not included within the scope of the Large Commercial and Passenger Yacht Codes are now routinely incorporated in the design of construction of Large Yachts, the REG YC includes a set of Common Annexes in order to ensure that these features are now subject to regulatory requirements. The pertinent points to note from a design and construction perspective are as follows:

4.4 (a) Battery Installations (Annex A)

The requirements are based on the text contained in the Maritime and Coastguard Agency's Marine Guidance Note, MGN 550 (M+F) entitled 'Electrical Installations - Guidance for Safe Design, Installation and Operation of Lithium Ion Batteries' and encourage the use of Risk Assessments, Hazard Identification Workshops or Failure Mode Effect Analysis tools to identify the risks associated with a particular cell chemistry and ensure that suitable safeguards are provided. Both the adverse effects of battery system failure upon the vessel, its ability to continue to operate in a safe manner and the risks associated with thermal runaway and cell fires require careful consideration.

4.4 (b) Fire Places (Annex D)

Guidelines for the installation of Fireplaces, utilising either Solid Fuel, Ethanol or LPG, Charcoal Ovens, External Fire Pits, Barbecues and Spit Roast for use on open deck areas have now been included. Again, the use of Hazard Identification Techniques is introduced to ensure that suitable safeguards are identified and provided.

4.4 (c) Occasional Workers (Annex E)

Further to the standard definitions for 'Seafarers' and 'Passengers', the Codes now introduce a third category of person known as an Occasional Worker. Such persons usually form that part of a Yachts complement who are primarily employed in a shore-based role but may be found on board working in service for the Owner or Charterer (i.e. a Nanny, Masseuse or Bodyguard).

Occasional Workers shall:

- Have no emergency duties and have undergone familiarisation training such that they should be self-sufficient in the event of an emergency. Whilst such persons are required to the trained to a level that will ensure their own safety on board, STCW basic training is not a mandatory requirement.
- Be at least 16 years of age, medically fit, have a contract of employment equivalent to that provided for under the MLC and be provided with MLC compliant accommodation.

4.4 (d) Storage of Large Quantities of Petrol in Fixed Tanks

Recognising the need to carry larger quantities of Petrol than the 150litre limitation in LY3, clear guidance concerning tank construction and location, passive and active fire safety measures, requirements for pumping and piping and operational procedures to be included in the Safety Management System are now included.

4.4 (e) Helicopter Landing Areas (Annex H)

In the interests of adopting an international standard, the Code now requires compliance with the applicable requirements of ICAO (the International Civil Aviation Organization) Annex 14. Matters relating to aviation are fully delegated to the Aviation Inspection Bodies (AIB's) who should be engaged at an early stage in the design development process. The scantlings of the deck are to be in accordance with the requirements of and approved by the Yachts Classification Society. Fire Safety measures are based on the requirements of SOLAS II-2/18 although modified to accommodate the use of Deck Integrated Fixed Firefighting Systems. A Helicopter Landing Area Technical Certificate is issued by the AIB one the verification process regarding the Design and Construction details is complete and will assist Builders with concluding their contractual obligations upon delivery. A Helicopter Landing Area Certificate may subsequently be issued once the operational requirements (i.e. Crew training and certification and the development and implementation of suitable procedures) have been implemented.

4.4 (f) Protection of Personnel (Annex O)

Cleary defined standards are now included for Overside Working Systems, Personnel Lifts and Man-riding Lifting Appliances.

5. KEY TECHNICAL CONSIDERATIONS DURING THE DESIGN AND CONSTRUCTION PROCESS

5.1 MINIMUM SAFE MANNING AND THE MLC

Common Annex G now includes the basic principles of Minimum Safe Manning in accordance with IMO Resolution A.1047(27). Clarification of those items that require careful consideration in order to ensure that the Yacht can be operated in a safe and secure manner is now provided. The importance of ensuring that the Flag Administration is engaged at an early stage in the design process to verify that Crew Accommodation arrangements are sufficient to satisfy the requirements of the Minimum Safe Manning Document cannot be over emphasised.

For Part A Yachts, the substantial equivalences to the ILO'MLC 2006 including the acceptance of twin cabins in Large Commercial Yachts of 3000GT to less than 5000GT for Seafarers who are not performing the duties of officers, which were originally published in MGN 517 (M) are now included in Chapter 21B.8(5). Part B Yachts continue to be subject to compliance with the MLC as applicable to Passenger Ships, requiring Officers to be provided with their own cabins. Those seafarers not performing the duties of ships officers may be accommodated in cabins of up to 4 persons.

5.2 MEANS OF ESCAPE AND EVACUATION

Code requirements for Part A Vessels regarding Means of Escape and Evacuation remain as per the previous requirements in LY3 except for Section 14B.13. The additional text introduces new requirements regarding the clear widths of stairways (including access doors) corridors and escape hatches in addition to requirements for the provision of handrails.

Code requirements for Part B Vessel remain as per the 6th Edition of the PYC / SOLAS II-2/13.3.2 (Means of Escape in Passenger Ships). During the early stages of developing the General Arrangement Plan, it is important to note that Part B 6.11 (13) requires the provision of at least one stairway enclosure in each Watertight Compartment contrary to the acceptance of limited vertical escape routes (i.e. via ladders and hatches), which may be accepted for Part A Vessels.

5.3 STANDARDS OF STABILITY & LSA SCALES

5.3 (a) Part A Yachts (greater than 85m Loadline Length)

Part A Vessels are required to comply with Intact Stability criteria taken from the IS Code 2008, Section 2.2 (criteria regarding righting lever properties). The

Multi-Hull Motor Yacht Criteria contained in LY3 is now referred to as 'Alternative Criteria' in Part A and may subsequently be applied to Mono Hulls where compliance with the standard criteria is not possible. Unrestricted vessels are required to comply with a one compartment standard whereby a minor damage methodology is adopted in which damage is assumed not to occur on any watertight boundary.

LY3 and the REG Yacht Code Part A require 100% Capacity on each side of the vessel. Two additional options are provided whereby Lifeboats may be substituted by Liferafts subject to compliance with SOLAS two compartment damaged survivability criteria. Alternatively, Lifeboats may be substituted by Davit Launched Liferafts. In such cases, one approved rescue boat shall be provided on each side of the vessel for the purpose of marshalling and towing duties.

Part A now includes an option to utilize a Marine Evacuation System (MES) as the sole means of abandonment. In such cases, at least one suitably sized inflatable slide or chute shall be provided on either side of the vessel. MES shall be capable of being deployed from the ship under unfavourable conditions of trim of up to 10° and list of up to 20° either way and in the case of being fitted with an inclined slide, operate such that the angle of the slide to the horizontal is within a range of 30° to 35° when the ship is upright and in the lightest sea-going condition; and a maximum of 55° in the final stage of flooding. MES are to be located no higher than the bulkhead deck and enclosed by fire rated divisions.

Furthermore, all Yachts irrespective of Gross Tonnage now require ship-specific plans and procedures for recovery of persons from the water, taking into account the guidelines contained in MSC.1/Circ.1447. The plans and procedures shall identify the equipment intended to be used for recovery purposes and measures to be taken to minimise the risk to shipboard personnel involved in recovery operations.

5.3 (b) Part B Yachts

Part B Vessels are required to comply with the IS Code 2008 as applicable to Passenger Ships. It should subsequently be noted that in addition to the General Criteria that is applicable to Part A Vessels, compliance with the Passenger Crowding and Severe Wind and Rolling Criterion is also required. Vessels having a Loadline length >80m and carrying Lifeboats must comply with Probabilistic Damaged Stability Criteria. Where Lifeboats are substituted for DLLR's, compliance with Deterministic Two Compartment Enhanced Survivability Criteria is also required. This subsequently results in a higher degree of watertight sub-division, than compliance with the probabilistic criteria requires. The benefits associated with the carriage of Lifeboats can therefore include a reduced requirement for the number of stairway enclosures or a larger beach club dependent upon how the watertight sub-division is arranged, [2] although any such benefits are off-set by the increased demand on open deck space that the Lifeboats and their associated Launching Appliances would require in comparison with DLLR's.

5.4 FIRE PROTECTION MATERIALS FOR PART B VESSELS:

The surfaces of interior linings, bulkheads, ceilings and floors for Part B vessels are harmonised with SOLAS Chapter II-2 Regulation 5.3 (Fire Protection Materials) regarding requirements for surface spread of flame, the production of smoke and toxic bi-products, combustible volume and calorific value. This constitutes a significant increase in regulatory requirements when compared with Part A Vessels, whereby any comparable restrictions for those vessels provided with an automatic sprinkler system and fully addressable fire detection system are very limited. It should however be noted that a great many pre-certified interior finishes are now widely available and luxury, bespoke interiors are being designed and constructed in full compliance with regulatory requirements.

Two methods of calculating Fire Loads are available, these being the fire volume method in accordance with SOLAS II-2/5.3 or the mass method in accordance with MSC/Circ.1003. Experience to date has indicated that the Fire Volume method is the preferred approach for interior designers and outfit companies although both methods are described here for the sake of completeness.

5.4 (a) Calculation of Fire Load - Volume Method

With respect to the calculation of fire loads, the maximum permissible Calorific Value of combustible materials used on bulkhead and ceiling linings shall not exceed 45MJ/m2 for the thickness of veneers used.

The total volume of combustible facings, mouldings, decorations and veneers in accommodation and service spaces shall not exceed a volume equivalent to 2.5 mm veneer on the combined area of the walls and ceiling linings.

The requirements are not applicable to the surfaces of furniture fixed to linings or bulkheads, which need not therefore be included in the calculations.

The calorific value is to be determined by fire testing in accordance with ISO 1716:2010 (Reaction to fire test for products – Determination of the gross heat of combustion). Alternatively, calorific values may be utilised from other credible sources such as the data tables published by the NFPA or another equivalent source.

5.4 (b) Calculation of Fire load - Mass Method: [3]

It is important to note that unlike the fire volume method, the total mass of combustible materials shall include floor coverings, construction materials such as cable insulation, plastic pipes, veneers and combustible materials permitted to be used under the relevant regulations of SOLAS chapter II-2 and outfitting which may be installed during the construction or provided by the Owner or Crew, including furniture and bedding components.

The total mass of combustible materials per unit area of the space (TCMPA) in kg/m^2 is calculated as:

$TCMPA = M_{CMBST} / A$

Where:
M_{CMBST} = Total mass of combustible materials of the space (kg)
A = Floor Area of Space (m$^{2)}$)

The maximum permissible values for TCMPA are as follows:

- Stairways, Corridors and Control Stations: 5 kg/m^2
- Accommodation Spaces: 35 kg/m^2
- Service Spaces surrounded by "A" class divisions 45 kg/ /m^2

5.4 (c) Use of Non-Certified Materials

In order to ensure that interior spaces can be designed in accordance with the expectations of the Owner, Part B contains a provision for the acceptance of non-certified materials in luxury areas outside of those spaces forming escape routes.

In all cases, such materials should be kept to an absolute minimum and the Administration should be consulted either directly or via the Vessels Classification Society regarding the feasibility of acceptance. A detailed proposal should be submitted, which will be considered on a case-by-case basis. However, in general terms, interior designers should consider the following:

- The use of non-certified material should where possible be avoided in way of doorways used as a means of escape and at the junctions of bulkheads to ceilings
- The percentage area limits for non-certified material on ceilings and the upper region of bulkheads should be more restrictive than the mid and lower regions of bulkheads
- The provision of suitable fire breaks in way of non-certified materials should be considered. Such fire breaks should consist of non-combustible material and will need to be substantial to be effective. A continuous and substantial non-combustible protrusion (i.e. a deep beam) may also serve as a fire break
- The fire characteristics of the non-certified material should be confirmed via testing in accordance with FTP Code requirements and will ultimately determine whether a material could be accepted.

6. CONCLUSION

The new REG YC has now been substantially updated in order to incorporate the latest requirements adopted by the underlying conventions and to further promote its international acceptance within the wider maritime community.

The increased use of 'Risk Based Design' and 'Alternative Design and Arrangements' (AD&A) provides a greater degree of flexibility and innovation for Designers and Builders. However, it must be ensured that the risks and limitations associated with this philosophy are properly understood in order to achieve a successful outcome.

Additional flexibility has been provided by adopting the use of internationally recognised standards such as ISO/DIN/BSI.

7. REFERENCES

[1] Maritime & Coastguard Agency, Notification to IMO, November 2018

[2] A. Meredith Hardy, S. Julien, J. Roy, G. Evans, B. Rial 'The New Generation of Superyachts – SOLAS or PYC?, HISWA Symposium, 2012

[3] IMO Maritime Safety Committee, Guidelines on a simplified calculation for the total amount of combustible materials per unit area in accommodation and service spaces, June 2001

8. AUTHORS BIOGRAPHIES

Julian Smith holds the position of Principal Surveyor, Global Safety & Compliance at the Cayman Islands Shipping Registry. He is responsible for both Plan Approval and Site Survey activities regarding the Certification of Yachts in accordance with the Large and Passenger Yacht Codes from both a New Construction and In-Service Perspective. Previously he spent almost 13 Years with Lloyds Register in employed in field based and plan approval roles.

Carlo Vinelli holds the position of Senior Surveyor at the Isle of Man Ship Registry having joined in July 2017. His career has seen him involved with many shipyard activities as an external advisor or project manager, mainly focusing his experience on large commercial yachts, most of them registered under Red Ensign Group.

ANALYSIS OF CHANGINGS IN MEGA YACHT GEOMETRICAL CHARACTERISTICS OF HULLS WITH CONSEQUENCES ON RESISTANCE AND STABILITY

V.Ruggiero , Department of Ingegneria, University of Messina, [ITALY]

SUMMARY

The aim of this work is to provide the results of the analysis of a series of yachts, with similar characteristics, built in Italy in the last 30 years, of which both tank tests and stability tests were available, in order to identify the presence of similar geometric characteristics or the presence of a systematic modification of geometric relationships, such as to allow the possible identification of a type of similar hulls. The identification of similar characteristics, obviously ascribable to tested and built units, to the satisfaction of the shipyard and the ship owner, can be of help to the preliminary choices of the designer.

NOMENCLATURE

[Symbol]	[Definition] [(Unit)]
C_F	Friction coefficient [-]
C_{RS}	Residual resistance coefficient on ship [-]
C_{FS}	Friction coefficient on the full scale ship [-]
C_{TS}	Total resistance coefficient on the full scale ship [-]
Fr	Froude Number [-]
L_{WL}	Length in the waterline [m]
V	Linear velocity [m/s]
V_S	Velocity of the ship [m/s]
B	Breadth, moulded [m]
B_{WL}	Breadth in the waterline [m]
L	Length in general [m]
L_{OA}	Length over all [m]
L_{WL}	Length in waterline [m]
L_{PP}	Length between perpendiculars [m]
S	Wetted surface [m^2]
T	Draught, moulded of ship [m]
C_{TL}	Telfer coefficient [-]
D	Height (main deck height from base line) [m]
LCG	Longitudinal position of center of gravity [m]
VCG	Vertical position of center of gravity [m]
GMT	Metacentric Height [m]
LCB	Longitudinal position of center of buoyancy [m]
VCB	Vertical position of center of buoyancy [m]

1. INTRODUCTION

The definition of Yacht, as known, comes from the use of the Dutch word "Jaght" in the 17th century, to indicate a type of boat particularly fast and manageable, but over the centuries, and in particular in the last decades, yachts have undergone very fast changes as for their sizes and purposes. Initially yachts were born as a transformation of existing ships, with a review of the interior [1]. The type of ship most similar as a destination of use, i.e. the transport of a limited number of people, was generally the patrol-type military vessel or a support vessel, with subsequent modifications. In addition, initially the yachts were mainly Government ships used for representation tasks while the private yachts had smaller dimensions.

However, the private yachts market has evolved very quickly: the size of yachts from the 60s to the 70s has tripled: in the 70s a yacht of about 30-35m overall was considered a large yacht, while today for many yards the 50m yachts are the normality; according to the latest data, the average size of yachts under construction in Italy exceeds 40 m. All this has led to researches carried out independently to identify the best type of hull, both from the point of view of resistance to motion and of stability. That's why the present work aims at giving give a more critical and systematic organization for the examination of these parameters.

Eleven yachts have been examined, built in Italy in the last 30 years, with construction and performance characteristics such as to make them comparable and the main dimensional parameters and their variations were analyzed, both in relation to years and dimensions. Similarly, the resistance coefficients and their variations have been considered, as shown by the tank tests at our disposal, in order to examine their progresses. As known, in the ship field there is no "ideal hull", the hull of a ship was born and will always be born with the need to reconcile different requests, both regulatory and functional. In the case of yachts there is an added aesthetic need, but it is certainly possible to identify a sum of common characteristics or trends.

It has been therefore considered that it could be of help to the designer to carry out a systematic analysis that would allow, similarly to what was done by other authors for other types of hull [2], to be able to predetermine a range of such values, typical of large yachts, to be used as initial inputs in the design phase.

A precise typology of hull forms has been taken into consideration, however broadly representative of most of the yachts built or under construction.

2. SELECTION OF THE YACHTS TO BE EXAMINED

The initial selection of the units to be examined has been based on the possibility to identify yachts that, although constructed in different periods by different shipyards (with different classification Registers, however offered a substantial analogy in terms of hull types and performances.
This despite, as already mentioned and as shown for instance in Fig. 1, there have been important dimensional variations due to market and design requirements. The size indicated in the figure of a 41m Yacht represented the main dimension under construction in 1999, while the 72m of 2019 are more representative of the current market trend.

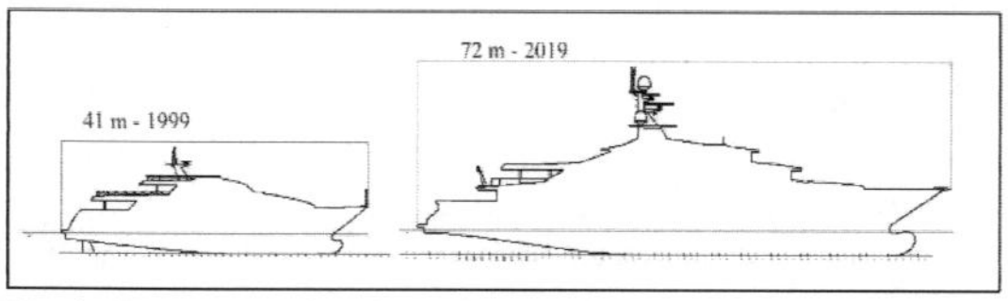

Fig.1 "Large yacht" in 1999 and 2019

The research is carried out in two successive steps: the first analysis is the one of the available types of yachts and the second is the analysis of the variation of the main geometric parameters with particular attention to the characteristic values of stability, such as the ratio between construction height D and height of the center of gravity VCG, between height metacentric GMT and ratio between the metacentric height GMT and the width of the vessel B. For all the examined ships the second analysis takes into consideration the examination of the tests in the model basin and the consequent identification of the coefficients of viscous resistance and residual resistance, and the total resistance values.
Furthermore, in order to try to provide a data of greater practical interest, the relationship between resistance to advancement and displacement at the test, the Telfer coefficient C_{TL} [3,4] and the ratio between the wet surface S and the water line lenght, L_{WL} ^ 2, have been highlighted to give an indication of the variation of the wet surface as a function of the variation of the dimensions. All the data under examination were treated through a normalization process, described later, in order to compare homogeneous conditions.
As already mentioned, eleven ships, whose data were available, were considered. It must be pointed out that the reference population of round bilge yachts, currently in the world, amounts to some hundreds of units. Among the considered units 5 exceed 60m in length, and currently there are about 200 private, non-governmental yachts over 60 m. total length in the world. Therefore the sample is considered to have its own scientific representativeness. All the ships in question have the same type of displacement hull, with absolute speed ranges that, although different, still allow to identify similar Froude numbers, comparable hull shapes (round bilge hull), bow with bulb and similar propulsion of two propellers type.

All the hulls are built in steel with an aluminum alloy superstructure, with a transom type stern and maneuverability ensured by arrangement with two rudders. This substantial morphological similarity has made it possible to hypothesize a homogeneous comparison which, despite the obvious difference not only in terms of length but also of volume (Fig.1) and therefore of Gross Tonnage, has allowed subsequently to extrapolate dimensionless parameters and average values of immediate interest in the designing phase. The following table (Tab.1) shows the main dimensions in length and the years of construction of the units examined:

Year	L_{OA}	L_{BP}	B	D	Displ. Light
	(m)	(m)	(m)	(m)	(t)
1989	49,5	42,8	9,24	5,4	360
1989	60	54	10,5	5	478
1990	63,5	57	11	5,2	760
1992	28,5	24,5	6,8	4,14	114,5
1999	41	34	8	4,5	202,4
2000	26,8	23,94	6	3,65	105
2001	32	27,2	7	4	140
2001	44,5	40	8,8	4,8	309,8
2006	42,65	36,2	8,2	4,5	296
2015	69,4	56,27	11,3	6	913
2018	72	58,27	11,3	6	1014

Tab.1 Main dimensions

3. ANALYSIS OF STABILITY CHARACTERISTICS

As already mentioned, all the units examined had a similar construction type; the analysis of stability characteristics is based on the identification of similar conditions, deduced from the "Stability booklet" and "Inclination report" files approved by the respective Classification registers for all the examined units. Of course the required stability conditions were similar for all the units, that met the parameters required by the intact ship SOLAS standard, (Fig.2)

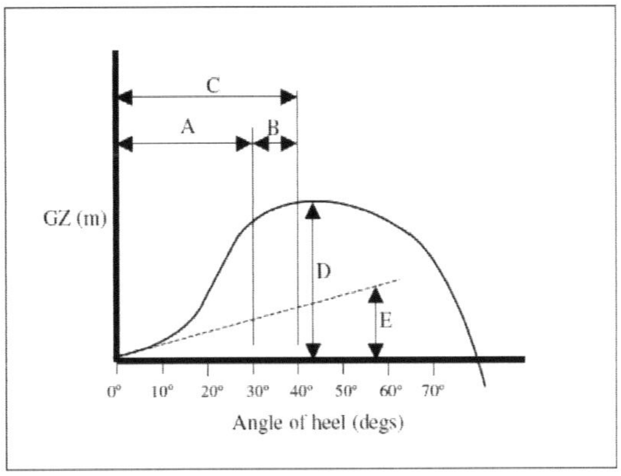

Fig. 2 RAH SOLAS Diagrams

Main SOLAS criteria:
A – Area under the curve up to 30° has to be not less than 0.055 metre-radians.
B - Area under the curve up to 40°, or the angle of down flooding, has to be not less than 0.090 metre-radians.
C - Area under the curve between 30° and 40° or the angle of down flooding (which ever is less), has to be not less than 0.030 metre-radians.
D - Maximum GZ is to occur at angles of heel preferably exceeding 30° but not less than 25° and the GZ value must be at least 0.20 metres at an angle of heel equal to or greater than 30°.
E - Initial GM is to be not less than 0.15 metres.

Once defined the similarity, the data have been collected, all referring to the same loading condition, taking into account the fact that all the ships examined had the Yacht classification, without exceeding the limit of 12 passengers. Referring to the Lightship reference condition with only passengers on board, it has been possible to analyze the values trends, in order to minimize any differences in stability values due to the use of Fuel or Fresh water as ballast.

3.1 VCG/D parameter analysis

One of the main parameters that the designer must face in the initial phase is that of estimating the vertical position of the VCG center of gravity: This estimate influences all the subsequent development of the project.
It is therefore appropriate to analyze the trend of this parameter over the years and its tendency. The graph in Fig. 3 shows how, in analogy with the increase of the Gross Tonnage and the tendency to increase the volumes of the superstructures, the ratio between the height of the center of gravity and the height of construction has been increasing during the years, regardless of the length of the ship and all this despite the use of ever lighter materials.

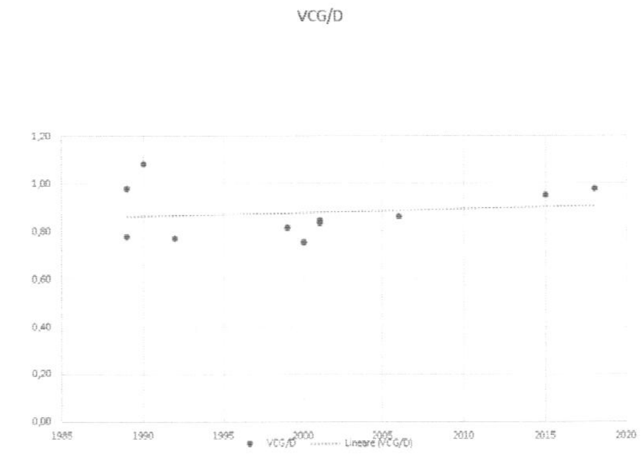

Fig.3 VCG/D trough years

However, the analysis of the values makes it possible to state that the VCG / D ratio tends to a value of around 85% of the construction height.

3.2 GMT/B and GMT parameter analysis

The GMT/B parameter (Fig.4) is not a classic parameter, as the geometric correlation between the height of the metacentre and the width of the ship is well known and it is not necessary to recall it here, but it has been considered appropriate to report this parameter relating to the years of construction of the various units examined, units different from those presented in other works [5], as it has been noted that, over the years, this parameter has undergone a rather significant variation. As it can be seen, it passes from a value of about 0.13 at the beginning of the period considered to a value of about 0.16, with a percentage increase of 20%.

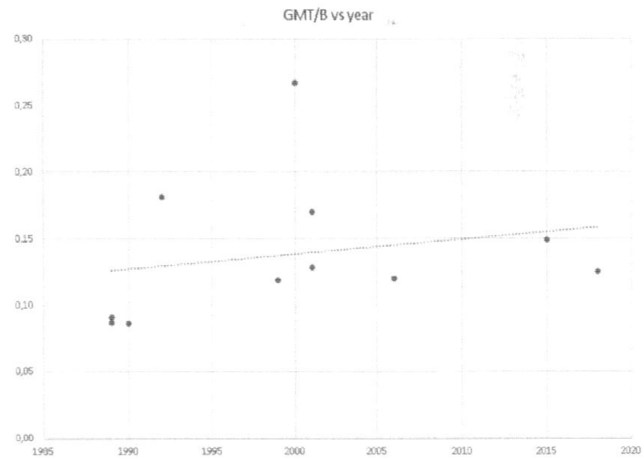

Fig. 4 – GMT/B trough the years

Such a significant increase is to be attributed to a variation in the shapes of the waterlines figure, but this value must then be correlated with the data concerning the resistance to motion, which will be examined in chapter N. 4.
By way of analysis, Fig.5 shows the value of the GMT parameter only and its variation. Also in this case it can be

seen that there is a definite tendency to increase the value of GMT.

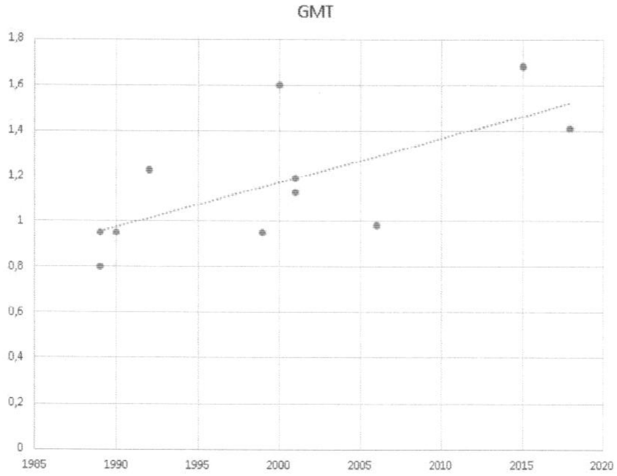

Fig.5 GMT variation trough the years

3.1 Analysis of ratio T/D

Fig. 6 shows the trend of the ratio T/D, i.e. Draft/Height over the years. It has been considered appropriate to highlight this ratio to have a parameter of the relationship between the variation of the immersion, possibly linked to a variation of the dimensions and a variation of the construction height. The substantially constant trend of this relationship over the years, allows us to assume an initial design value that can be used, once the displacement and then the immersion are known, as an indicator of the freeboard necessary to comply with the provisions on flooding angles and as a retroactive verification of the vertical position of the center of gravity.

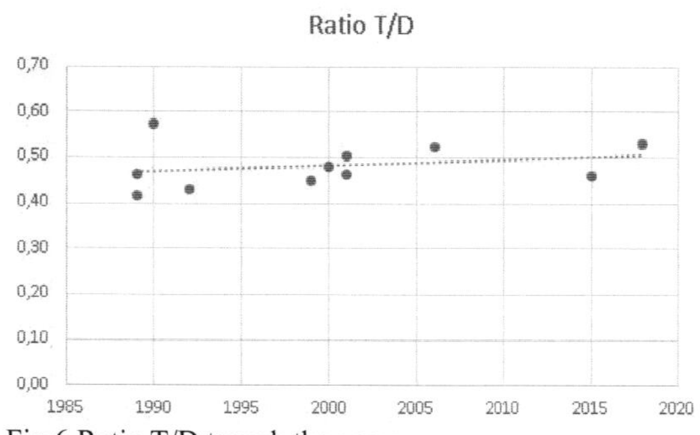

Fig.6 Ratio T/D trough the years

4. ANALYSIS OF HULL RESISTANCE DATA.

The analysis of the characteristics of resistance to motion, based on the same hulls used for the identification of the geometric parameters related to stability, has required, by its nature, a greater work of data normalization. The starting point has been the analysis of the tank tests carried out at the time of the construction of the ships. Obviously, since these tests were carried out in institutions belonging to the ITTC (International Towing Tank Conference), the verification of the uniformity of the test criteria (ie the use of the ITTC 78 methodology with presentation of results with reference to standard values) has been immediate.

For all the hulls examined (despite the similarity of the propulsion system and Appendages Plan layout, therefore two-propellers configuration, two rudders, transom stern, bow thruster, anti-roll active stabilizers) it has been decided to consider the resistance values resulting from the bare hull resistance tests, without appendages. All the tests considered are those related to the full load displacement. It has been verified, by comparison with the approved Stability booklet, that the full load displacement adopted for the tank tests coincided with the actual displacement during the operation of the vessel, in order to avoid to base the research on data, scientifically precise, but of a purely theoretical validity and not corresponding to the reality.

Obviously, speaking of motion resistance, it has been necessary to make the data comparable from the speed/length point of view. Therefore the Froude numbers of the hulls in question have been examined, referring to cruising speed: the hulls, with water lengths included between 25 m and 62 m, had values of the maximum speed between 14.5 and 20 knots. It has been seen that the values of the required cruising speed, between about 9.5 and 15 knots have identified a variation of the number of Froude Fr. between 0.31 and 0.34 with a concentration of values around 0.33. Therefore a Froude number equal to 0.33 has been assumed as the reference value for all the hulls and the values of the total resistance RTS and of the resistance coefficients at this speed have been obtained, obviously using the values referred to the ship and not to the model, and therefore C_{FS}, C_{TS} and C_{RS} have been derived.

Without prejudice to the obvious theoretical discussion, but here wishing to privilege the achievement of indicators of greater practical use, with respect to a theoretical treatment that is in any case widely known, the value of the wet surface S, and the ratio Total resistance ship/Displacement, have been calculated leaving it expressed in kN /t, for a greater immediacy.

According to what already mentioned in the points by Telfer and Van Oossanen et al. [3] and [4], in order to make the hulls equal from the resistance point of view, the Telfer C_{TL} coefficient has been obtained. This coefficent is known to be defined by the following equation.

$$C_{TL} = \frac{RT/\Delta}{\left(\frac{V}{\sqrt{LWL}}\right)^2} \text{ or}$$

$$C_{TL} = (R_T{*}L_{WL}) / (\Delta * V^2).$$

From the examination of the data thus collected it is therefore possible to obtain a general trend of the resistance coefficients.

4.1 Analysis of Telfer coefficient

The Telfer coefficient for all the hulls considered, at a Fr = 0.33 has been calculated for all the hulls, and the graph in Fig. 7 shows that trough the years the coefficient had a little increase.

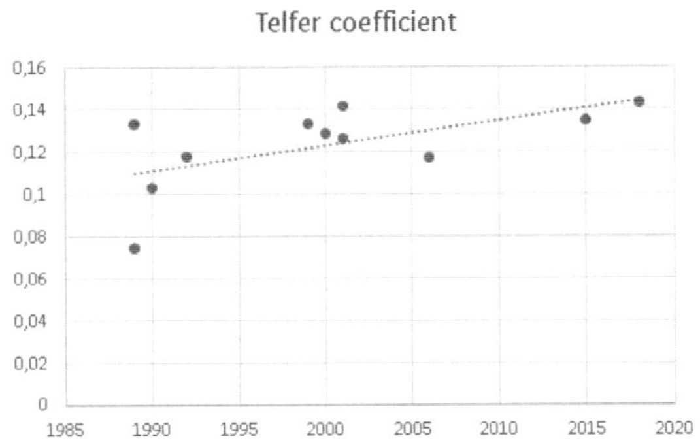

Fig.7 Telfer coefficient trough years

But the trend of the Telfer coefficient as a function of length is more interesting: with the same Froude number, it shows an improvement in increasing the ship length. Fig. 8

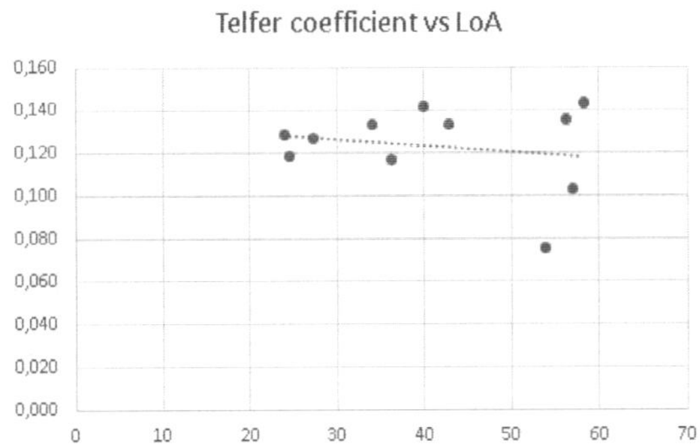

Fig.8 Telfer coefficient vs L_{OA}

4.2 Analysis of ratio wet S/ L_{OA}^2

A further examined relationship is the one that, at the test conditions, considers the ratio between the wetted surface and the waterline length of each hull, (Fig.9) squared for dimensional reasons. The graph , for obvious reasons reports on the X-axys the length. This parameter can be used as an indicator of the ratio between the development of the effective surface of the hull, and the length, and can therefore be used as an indication of the total final resistance values, if one wishes to adopt the classic ITTC methodology starting from an estimate of the Coefficients of resistance.

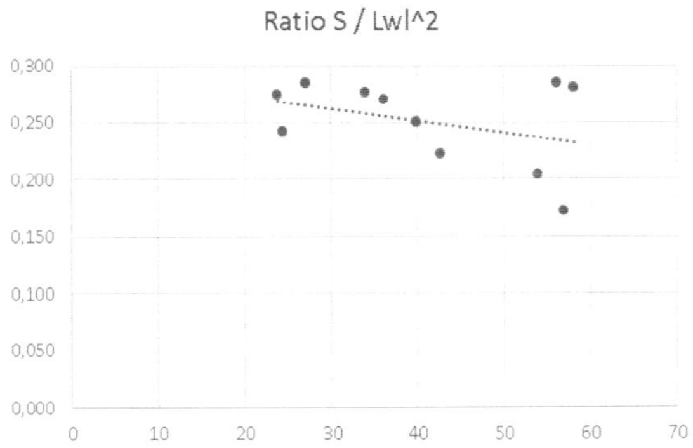

Fig. 9 – Ratio S/L_{WL}^2

4.3 Analysis of coefficient C_{RS}

The coefficient examined has been the Residual resistance coefficient C_{RS}, obtained from the tank test, in this case the variation trough the years (Fig.10)appears more as a constant value, but it is still important to have a value to be used as a preliminary input data at the corresponding conditions.

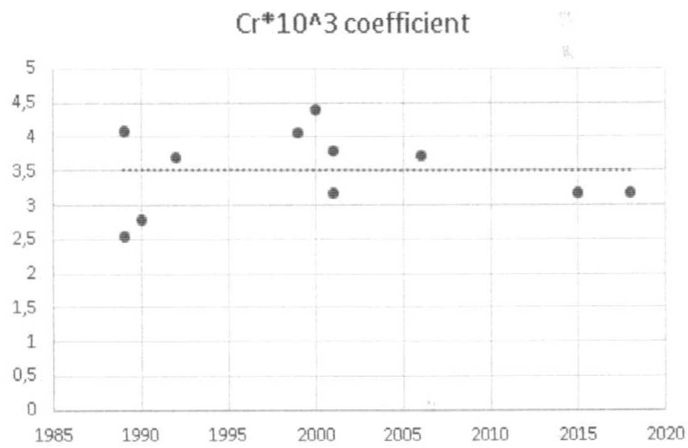

Fig.10 C_{RS}

5. CONCLUSIONS

As known, ship design takes place by means of successive steps, in which one or several parameters are refined as a result of the choices made, to return to examine the global product and introducing successive optimizations, according to the well-known axiom defined as "project spiral".
Generally in this preliminary phase some characteristic parameters are chosen either from bibliographic research or simply based on a "Model" vessel which is then modified by the designer to adapt it to the contractual specifications. In this phase it is evident the utility that the knowledge of a series of average values can have to tend to or from which to depart, also evaluating if the model ship re-enters or not in these parameters, in order to be able to minimize the iterations of optimization. The analysis of the selected hulls has shown the tendencies of the variation

of a series of parameters for yachts actually realized in an arc of 3 decades and all still navigating.
We can therefore easily extrapolate from the results obtained, which are only some of all those obtained by the series of tests and documents examined, some average values to be taken as project input values. The preliminary definition of these values, for the type of examined hulls, in any case is considered representative of most existing yachts or under construction.
Summarizing: the analysis of the hulls examined allows to indicate the average values for the following parameters:

Parameters	Average values
VCG/D	0.83 - 0.86
GMT/B	0.13 - 0.16
GMT	1-1.5 (m)
T/D	0.47 – 0.5
C_{TL}	0.11 – 0.141
S/L_{WL}^2	0.23 – 0.26
C_{RS} * 1000	3,49 – 3,15

Tab.2 Average values to be adopted

6. ACKNOWLEDGEMENTS

Author wants thanks the shipyards, designers, and Tank facilities who gave availability of the data for the present research.

7. REFERENCES

[1] RUGGIERO,V. (2016) "Changes in design approach for large yachts"
SGEM 2016, Vienna ,(Vol.4 Pages: 155-163)

[2] BAILEY, D., (1976) "The NPL high speed round bilge displacement hull series", Maritime Technology Monograph No. 4., Royal Institution of Naval Architects, 1976.

[3] TELFER E.V., (1963) "The design presentation of Ship model resistance data", North east coast institution of Engineers and Shipbuilders. Transactions Vol.79,1963.

[4] VAN OOSSANEN P., HEIMANN J., HENRICKS J., HOCHKIRCH K., (2009) "Motor yacht hull form design for the displacement to semi displacement speed range" , 10th International Conference on Fast Sea Transportation, FAST 2009, Athens, Greece, October 2009.

[5] RUGGIERO,V. (2018) "Systematic Analysis of geometrical characteristics of hulls for displacement yachts ",June 2018 Conference: NAV 2018 - 19th International conference on ship & maritime research,DOI: 10.3233/978-1-61499-870-9-845

8. AUTHORS BIOGRAPHY

Valerio Ruggiero (Ph.D) is actually Researcher at University of Messina, Italy, where he is Professor of courses of Naval Hydrostatics and Naval Architecture. He is responsible of research activities in the field of Tank Tests and CFD , with special regards to ferries, hydrofoils, fast crafts.
His previous experiences includes 15 years of work as Naval Architect in cohoperation with Italian and international shipyards for realization of ships like yachts, ferries, drydocks, patrol vessels.

OUT-OF-THE-BOX INTEGRATED, COLLABORATIVE, MULTI-AUTHORING, MANAGED ENVIROMENT FOR THE DESIGN AND CONSTRUCTION OF LARGE YACHTS

N W Danese and **S Dardel**, Nick Danese Applied Research, France

SUMMARY

Modern computational technology available to the large yacht industry spans Naval Architecture and ship building disciplines but only rarely are tools used synergistically and processes connected, let alone integrated. Nonetheless, an Out-Of-The-Box Integrated, Collaborative, Multi-Authoring Environment for the Design and Construction of Large Yachts composed by a heterogeneous software pool spanning 3D modelling, weight engineering, CFD, structural analysis & fatigue, seakeeping, stability, CAD, etc. is available, today. The corresponding non-linear work flow is defined for practical application with no requirements for specialist training or specialized staff past Naval Architecture education for some and following LEAN and AGILE business models and strategies. Commonly used software tools are employed across the enterprise's landscape to cover all salient activities and processes. A *quasi*-PLM perspective is made possible by a structured information management process and the creation and maintenance of unique information models which feed the Work In Progress and Contractual deliverable archives.

NOMENCLATURE

3dm	3D graphics file format developed by the openNURBS Initiative
AGILE	method of project management that divides tasks into short phases and frequently reassess and adapts plans
AR	Augmented Reality
BOA	Beam OverAll (meters)
BOM	Bill of Materials
BPA	Business Process Assessment
COTS	Commercial Off The Shelf
D	Depth, the height of the main deck above the keel (metyers)
d&i	the data and information continuum
DDROM	Database Driven Relational Object Model
DISPL	Displacement (metric tons)
DMS	Document Management System
DWG	file format used for storing 2D and 3D geometry and metadata, used by Autodesk and other software houses
ECO	Engineering Change Order
EP	EnterprisePlatform, an SSI, Canada product
ERP	Enterprise Resource Planning
FTP	File Transfer Protocol
GA	General Arrangement
HRM	Human Resource Management
KPI	Key Performance Indicator
IC	Implementation Consulting
IIoT	Industrial Internet of Things
IIIoT&S	Intelligent Industrial Internet of Things & Services
LEAN	Lean manufacturing involves never ending efforts to eliminate or reduce waste or any activity that does not add value to the product
LOA	Length Over All (meters)
MCO	Manufacturing Change Order
MIM	Marine Information Model, the database underlying the ShipConstructor environment and intelligent model
ODBC	Open DataBase Connectivity
PLM	Product Lifecycle Management
ROI	return on investment
SSI	ShipConstructor Software Incorporated
STL	file format for stereolithography used in CAD to describe perfectly closed volumes
T	draft (meters)
USB	Universal Serial Bus supports all sorts of data transfer via a reliable, portable hardware connection
VPN	Virtual Private Network
VR	Virtual Reality
XML	eXtensible Markup Language
WAN	Wide Area Network

1. INTRODUCTION

Requirements and constraints imposed on every industry are ever harder to satisfy from commercial perspective. Response and adaptation vary greatly across the large yacht industry and a working PLM implementation is attempted by larger companies.

In fairness, the overall *concept-to-delivery* process behind the delivery of large yachts is arguably the most complex of all, comparable the space and the sub-atomic research industries. Moreover, large yachts include an element of exclusivity and luxury which, coupled with size, is a unique factor determining cost and customer satisfaction. Design work and manufact must be presented and satisfy these constraints, accordingly, supported by renderings, VR, AR and IIoT in addition to performing excellently from a Naval Architecture and engineering perspective.

Software available to designers and builders of large yachts spans the complete Naval Architecture and ship building horizon but only on rare occasions are the tools and the processes connected amongst themselves and with each other synergistically.

In contrast and perhaps ironically, most of the software tools so commonly used that they have become *de-facto* standards are very well suited to compose a highly productive framework *out-of-the-box* if operated in a shared, collaborative and managed fashion.

2. REQUIREMENTS AND CONSTRAINTS: AN OPPORTUNITY

Requirements on the marine industry are similar to those on other commercial operations, the most apparent ones including:

- faster
- cheaper
- more product

to which it might prudent to add for sake of financial success and company longevity:

- higher margins
- high customer satisfaction
- lowest post-delivery cost
- system and process renewal and evolution

and specifically, for the large yacht industry:

- the yacht's is evidently exclusive and permeated with luxury

The first set of requirements might feel overwhelming, but the ensuing forced acceleration of all industrial processes is a precious opportunity to make better use of existing resources to create a more productive environment with remarkably little investment. The software tools required are available and mostly already in use, the investment is hence limited to improvement of processes and of the human contribution. Both will generate immediate ROI at all levels and, especially, at the company level. Business Process Assessments, Implementation Consulting, hands-on training, etc. are commonly available and obvious means to this end.

3 CURRENT DESIGN AND PRODUCTION ENGINEERING ENVIRONMENT AND PRACTICES

Most companies in the large yacht industry design and production sectors are generally equipped with modern software and sometimes industrial tools (such as ERP platforms and robots). Most modern software is ready *out-of-the-box* to compose and support an Integrated, Collaborative, Multi-Authoring, Managed Environment, but few implement the framework and processes needed to turn that potential into ROI.

Crucially, the vital exclusivity and luxury components of the process stand to benefit, too, from a high-performance industrial process that yields a better product at a lower cost, thereby increasing the financial margin of the shipyard.

3.1 PERCEPTION AND LEGACY

The reasons for the above are many and varied, spanning from the culture of suppliers (designers, shipyards, etc.) to the expectations of customers (shipyards, yacht owners, etc.), and extend to include macro-factors such as political, social, and economic contexts, too.

3.1 (a) Perception

Culture is less of an obstacle than widely thought. Skilled, creative craftsmen are found world-wide, as are industries using advanced technology and producing superior quality products.

A facilitating reality is that the software tools and processes required to create a more productive and higher ROI environment are known, available and already used globally, if perhaps in different ways. This effectively extends the reach of the environment discussed here.

3.1 (b) Legacy

Several different approaches to achieve process performance are in use across the large yacht industry but, interestingly, none seem to be prevalent. To generalize a little, some clustering of approaches to performance can be attributed to culture and values shared in a socio-geographical context, but the personal, individual component appears to be predominant.

On the other hand, the speed at which technology has progressed since the start of its democratization and vulgarizing less than two generations ago is an objective cause for many managers to be unaware of the potential benefit intrinsic in the wider application of commonly found IT tools - some of which they themselves use daily. LEAN and AGILE business models are largely unknown to the large yacht industry, too.

Not least, "consumers" and "sources" of data are still too commonly identified with people, while machines and processes use and generate data and information, too, and increasingly more so than humans.

3.2 MARINE 4.0

The majority of companies in the yacht industry appear to be in the middle-ground between underperforming and highly effective operations. Although data sharing, robots and IoT are found in some instances, when referring to the currently common Industry / Marine 4.0 yardstick, many designer, shipyards and supply chain companies straddle the 2.0 and 3.0 spaces.

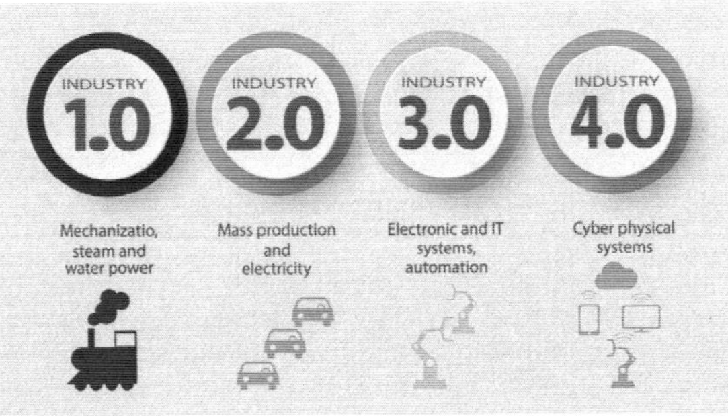

Figure 1: Phases of industrial evolution

One root cause of this is the lack, if not absence, of a managed data & information sharing policy and strategy accompanied by an under-exploitation of tools already in common use. Integration, cooperation, sharing, ERP and PLM are as commonly talked about and referred to as they are not implemented effectively, or at all.

3.3 STATUS-QUO

Arguably, little has changed in data and information sharing techniques, processes and strategies in the yacht and in many cases even in the large yacht industries since the mainstream CAD and spreadsheets became available on the first PCs in the early-1980s, followed by user-friendlier databases in the early 1990s. AutoCAD® (by Autodesk®, USA), Microsoft® Excel® and Access® continue to be used and at least the former two remain the *de-facto* ubiquitous backbone of most companies.

Perhaps ironically, AutoCAD v.1 was a full-fledged 3D CAD program in 1982, Excel always offered a multi-dimensional matrix and weakly relational environment, Access was released in 1992 as a fully relational database system and all were including interfacing and integrating capabilities since the mid-1990s.

Another current world-wide CAD *de-facto* standard is Rhinoceros®, by McNeel & Associates, USA, better known as Rhino3D®. Its 1997 beta version adopted by 100000 users within one year, Rhino3D's filled a technological void with user-friendly surface modelling.

Victims of their own *ease-of-use*, these and other common software are used by virtually everyone to satisfy a limited, immediate need but effectively exploited only marginally and at a constantly diminishing fraction of their growing potential. This represents substantial user underperformance and a considerable missed revenue opportunity at the company level.

The lack of investment in implementing and exploiting existing tools for overall company ROI is not the sole area where short-sightedness is the cause of false economies. For example, man hours are often counted as such, e.g. a straight cost, as opposed to being valued and assigned as a function of their potential to generate added value. Current practice still includes reliance on:

- (subjective) human communications
- assumption that messages, duties, tasks, deliverables, etc. are evident, clear and known to everyone
- cost minimization takes precedence on ROI-producing choices

4 GENERATION OF ADDED VALUE

Added value is vital at the company level, that is how much was spent *vs* the sold-at price, an often-ignored metric. On the other hand, the potentially misleading measure of *perceived* "margin" at the microscopic level, typically the number of hours taken to produce something, takes precedence leading to potentially catastrophic false economies.

Immediate opportunities to generate added value are the reduction, if not elimination, of errors, multiple instances of data and information and confusion.

4.1 ERRORS, MULTIPLE INSTANCES OF DATA AND INFORMATION

Errors force re-work, which in itself causes a domino-cascade effect of resource reassignment and compounded losses. And, the farther along the way an error occurs, the greater the damage.

Multiple instances of data and information are a genetic, self-immune disease of every process, and create noise and confusion throughout the framework of the design & production space. The unavoidable incompleteness, skew and subjectivity of each instance adds to the uncertainty, which people attempt to palliate with repeated data transmission effectively increasing confusion and reducing control on outcome.

Added value and ROI are generated immediately upon taking remedial action.

4.1 (a) Unforced Errors

The sports-world expression "unforced error" fits our industry well. Most errors are not due to objective reasons and should not happen. Typical causes include:

- use of insufficient, inaccurate data
- absence of a contingency plan
- no or bad communications
- untimely decisions
- lack of implication of stakeholders in the process

- data and information are not made available
- unmanaged data and information environments
- absence of data and data tracking via internal deliverables
- absence of structured workflow, planning and managed performance tracking
- insufficient advance vetting of input and resources
- etc.

but also, and key to this discussion

- heterogenous formatting of data requiring conversion and causing quality deterioration
- data formatted inadequately for its use
- impossibility to retrieve the information being passed due to the formatting of the data, its inaccessibility, etc.
- etc.

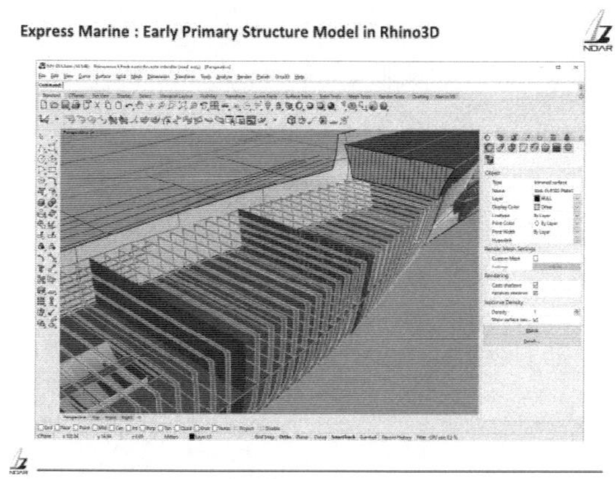

Figure 2 : Express Marine Very Early Primary Structure Model of a Super Motor Yacht in Rhino3D

4.1 (b) Multiple Instances of Data and Information

This condition is as common is the of software such as AutoCAD, Excel, etc., the perverse use of email and Windows® Explorer® as data, information and file management systems. The absence of data and information management policies (ex. even just naming conventions) and strategies at the company level does not help, and specialized software such a DMSs is rarely implemented despite its common availability.

Other causes include:

- unjustifiably heterogeneous software environments
- lack of standards
- individual preferences overruling overall company policy and best interest
- etc.

4.2 UNUSED AND UNKONWN DATA

The amount of unused data in the project space is considerable and often exceeds the data represented in drawings which often are, at best, subjective, arbitrary and incomplete representations of an unmanaged data set. Crucially, unused data includes *up-to-date* data that is not exposed, data that is not known to exist, etc. In the absence of a structured, managed approach to data management as proposed in this study, the legacy reliance on drawings compounds itself into is a major cause of error.

To use and expose *all* data and information, up-to-date and custom-formatted, ready-to-use by each stakeholder is the single most important factor in the success of the proposed approach.

5 COMMUNICATIONS

Evidently, communications rank high within crucial factors instrumental in achieving an effective, high ROI operation. Structured and managed communication policies, schemes and strategies are one of the DNA strands of the Integrated, Collaborative, Multi-Authoring, Managed Environment.

Reliance on *people telling people* is a guaranteed cause of communication failure. Many weaknesses ballast down this approach:

- timing: often off, planning not respected, . . .
- content: incomplete, not applicable, . . .
- reliability: validity/maturity of the data, accuracy, . .
- clarity: ready for use by the recipient, . . .
- completeness
- destination: did the prescribed recipient receive the required data, . . .
- traceability: where did the data come from, . . .
- etc.

5.1 (a) Available vs Accessible

Most data, information and goods could be *accessible*, if one knew where to look, how to get them, the format being adequate for this or that consumer, etc. But better still data, information and goods can actually be made *available*, that is delivered to this or that consumer, at specified times, in the appropriate format, etc., very much in LEAN fashion.

Unique Models (essentially data sets) compound the power and effectiveness of making data, information and goods available. A Unique Model simply indicates that what is being made available is specific to the recipient in its contents, formatting, timing, etc.

The Google® *vs* Facebook® analogy illustrates this.

A successful Google search requires that the target is present on the internet, that the question allows the Artificial Intelligence algorithms to converge (try "image of man with mammoth"), and that (if it exists) it is displayed in the first one or two pages.

Facebook is a *"defined"* subscription service: the consumer specifies exactly what is to be sent - pictures, posts, etc. posted by specific people - and these are delivered directly to the consumer's desktop as soon as they become available.

5.1 (b) Shared, Collaborative Environments

First and foremost, it must be recognized that the greater the number of stakeholders that share and collaborate, the greater the added value generated. Stakeholders include everyone and every company directly or indirectly involved in:

- design
- supply of goods, services, etc.
- production
- logistics
- planning
- etc.

A successful communication environment rests therefore on several pillars:

- data and information pertinent to the project exists in unique occurrences, everything is present only once in the project space, one-way post-processed for downstream consumption
- clear consumer's definition of required data, information, format and time frame of delivery
- ability to produce such data and information
- ability to dynamically modify the above
- required data and information is indeed available to all
- shared, collaborative environment collecting the above
- notification strategy and system
- feedback loop
- participation of all
- etc.

Effectively, the above collects job specifications, deliverables, ECOs, MCOs, ERP, DMSs, HRM, planning, reporting, KPIs, etc., in other words the whole design, production and operation space under a "communication" PLM-type umbrella.

5.1 (c) PLM

Born in the mid-1980s at American Motors Corporation, PLM systems are offered today by several software houses: ARAS, SAP, Intergraph, Dassault Systèmes, Siemens, INFOR, Oracle, Parametric Technologies, etc.

But, the all-collecting, ubiquitous, pervasive, etc. world of PLM has proven a challenge to implement in our industry characterized by the presence of very complex and interconnected processes involving a large number of assemblies, made of one-off assembled components they themselves made of a large number of heterogenous parts. Come subjective factors are involved, too:

- insufficient identification of the processes to implement
- unrealistic implementation goals
- incompatible tool sets and processes
- difficulty in mastering marine industry requisites by PLM vendors
- very long times to Milestone 1, by when original assumptions may have become obsolete
- etc.

While certainly not a substitute for a correct PLM implementation, the ensemble of many commonly available and already in use software can be exploited in PLM-style to generate ROI immediately.

6 SOFTWARE

Software is a preponderant source of data and information. Moreover, data generates information, and vice-versa. The two being effectively one, they will be hereafter referred to collectively as "d&i".

Communications being a *sine-qua-non* criterion and requirement underpinning this study, some software categories are defined accordingly:

- general purpose platforms
- wholly Read/Write different native formats
- multi-discipline
- integrated: runs inside another software
- interfaced: shares d&i directly or via a native d&i format with no degradation
- directly compliant: Big Data, Digital Twin, AR, VR, IIoT&S: direct connection

Additionally, connected programs categorized as:

- share d&i indirectly via a file format transformation/conversion, degradation is not ruled out

are taken into consideration only if they contribute exceptionally to the overall process despite their connectivity limitations.

Other discriminating aspects include:

- size and industry-wide distribution of current user base

- options provided by the tool to communicate with all consumers: people, machines, other software, processes, etc.
- scope of application, e.g. design and production
- current and projected potential to serve the Integrated, Collaborative, Multi-Authoring, Managed Environment

A further level of categorizing is that of "purpose", an admittedly anachronistic approach given the more recent evolutions in just about all software, hence the quotes subtending an inaccurate category name, but nonetheless informative enough to justify being carried out.

- "CAD"
- d&i "management"
- file & document management systems
- rising technology: VR, AR, Big Data, Digital Twin, IIIoT&S

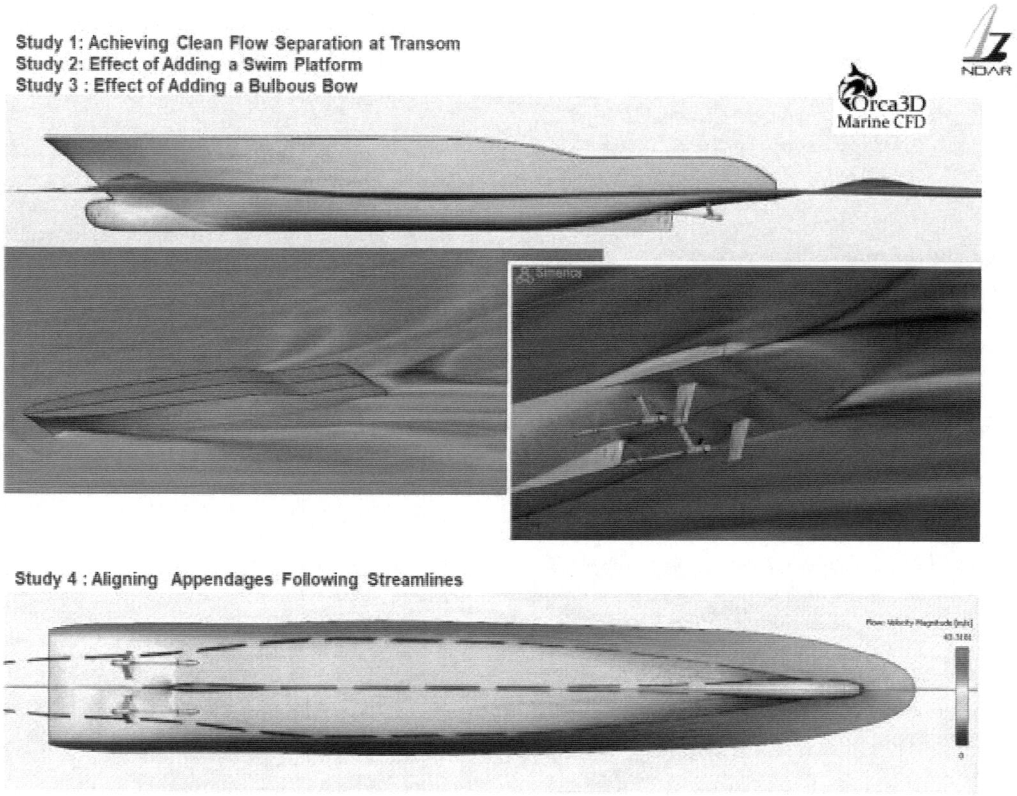

Figure 3: various CFD analyses on the Rhino3D model

6.1 (a) "CAD"

The primary goal of this research is the construction of an *out-of-the-box* Integrated, Collaborative, Multi-Authoring, Managed Environment made of tools that fulfil as many of the criteria listed here as possible. The Autodesk/AutoCAD and Rhinoceros platforms do so and additionally offer a number of unique business values, including:

- full compatibility at the native data format level with each other and several other general purpose and specialist software
- easy to use, short time to proficiency
- significant untapped productivity potential

The sheer number of candidates for the environment in object lends additional legitimacy to the findings of this study. An exhaustive list is beyond the scope of this study at this time, but a few representative products are mentioned for sake of documentation (in alphabetical order):

- 3DS StudioMax®, by Autodesk, USA
- Alleantia IoT-SCADA IIIoT&S, by Alleantia, Italy
- AdvancedSteel® by Autodesk, USA
- Ansys® by Ansys Inc., USA
- AutoFEM® by AutoFEM Software, Russia

- AutoShip® and AutoHydro® by AutoShip Systems, Canada
- Bongo® by McNeel Associates, USA
- Brazil® by McNeel Associates, USA
- Cadmatic®, by Cadmatic, Finland
- CMPIC, by Cloudis, GB
- CostFact, by CostFact, Germany
- DeltShip, by Delftship Maritime Software, The Netherlands
- Desys VR/AR, by Desys, Italy
- DropBox, by DropBox, USA
- Exchange®, by Microsoft, USA
- GHS® by Creative Systems Inc., USA
- ExpressMarine®, by BAS Engineering, Norway
- FORAN®, by SENER, Spain
- iSuite, by iYaldi, France
- Inventor® by Autodesk, USA
- MAESTRO® by Maestro Marine, USA
- NAPA® by NAPA Oy, Finland
- Nastran (various flavours by various editors)
- NavCad® by Hydrocomp Inc. USA
- Navisworks® by Autodesk, USA
- OneDrive, by Microsoft, USA
- Office®, by Microsoft, USA
- Orca3D® by Orca3D LLC, USA
- Orca3D Marine CFD® by Orca3D LLC and SIMERICS, USA
- Outlook®, by Microsoft, USA
- Penguin® by McNeel Associates, USA
- PIAS / Fairway® by SARC, The Netherlands
- Project®, by Microsoft, USA
- PropElements®, by Hydrocomp Inc., USA
- Revit® by Autodesk, USA
- RhinoCFD®, by Cham, Spain
- RhinoNest®, by TDM Solutions, Spain
- ShipConstructor® by SSI, Canada
- Solidworks® by Dassault Systèmes, France
- V-Ray, by Chaos Group, Bulgaria
- etc.

A core, short group of tools was retained for the *proof-of-concept* case study by referring most strictly to the adopted categories and discriminating factors. The author's proficiency in the use of certain tools was also considered.

The main mission of the shortlisted tools is summarized, and their synergy and interaction illustrated in map form elsewhere in this paper.

6.1 (b) d&i "management"

Managed, open architecture, API-equipped, transparently implemented, non-proprietary, databases are traits of effective d&i "management" in the present context. The following satisfy the brief:

- EnterprisePlatform® by SSI, Canada
- CostFact®, by CostFact, Germany

6.1 (c) File & Document Management Systems

Several mainstream software exist, for example:

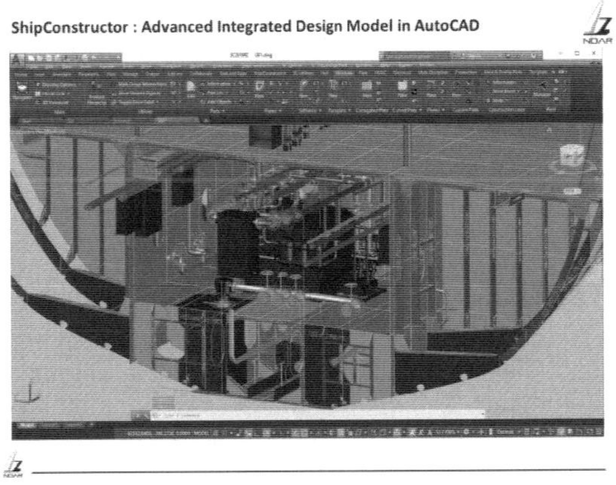

Figure 4: ShipConstructor integrated model, bow section of a large sailing yacht in AutoCAD.

- Meridian® by BluCielo, The Netherlands
- ProjectWise ®, by Bentley, USA
- M-Files, by M-Files, Finland
- etc.

However, the use of DMSs in general being more an exception that the rule prompts the search for other players. In this respect, a young tech start-up, iYaldi, France, recently released:

- iFileMinder
- iNotifika for EnterprisePlatform

which provide respectively file "management" with file presence/ absence notification in a shared, collaborative space and a supervision umbrella for SSI's EnterprisePlatform, complete with targeted, managed notifications of Operation execution and outcome. While not replacing a canonical DMS, iYaldi's tools support essential communication.

6.1 (d) Rising Technology: VR, AR, IIIoT&S

VR and AR are considered here as in their company-wide and future-compliant application, supporting an intelligent environment based on a model which includes:

- geometry
- meta-data germane to the "CAD" model but also coming from various sources (ERP, PLM, etc.)
- meta-data and field data from IIIoT
- compliance with the supply of Services in lieu of products

The "CAD" tools must therefore also be compliant with IIIoT&S. Among others, the DDROM data structure at the heart of SSI's MIM (the SQL database backbone of ShipConstructor, "managed" by SSI's EnterprisePlatform qualifies *out-of-the-box*.

7 A NEW PARADIGM

The fundamental requirements, multiple platforms, rising technologies, varied formats specific to the candidate software components and the diversity of consumers in nature, scope and purpose suggest a different approach to d&i use, storage and management, not dissimilar from the internet and processed Big Data models.

Moreover, consumers are in fact also a source of d&i via the feedback loop and the very outcome of their actions. The combined consumer-source entity is henceforth identified as an "*author* ".

The functional paradigm that emerges from this study identifies a "space" in which multiple authors evolve in a collectively non-linear, individually linear, multi-relational, multi-platform, multi-format, collaborative fashion. The "space" collects a variety of d&i in different, heterogeneous formats:

- text files
- proprietary format files (3dm, dwg, etc.)
- xml files
- databases
- etc.

Files are saved directly by the author in the designated native format, and in non-native formats unless absolutely indispensable as this undermines the integrity of the overall process.

The requirement for the space to be collaborative is evident, as is the need for file management and *state-of-* d&i notification.

It is worth noting that SSI's EnterprisePlatform presents *autonomous* ubiquity of service and document handling automation characteristics which allows a vast d&i set that extends far beyond what would be manageable with everyday manual procedures. The project space's archive grows to include crucial documents hitherto never considered, making EP one of the few *sine-qua-non* DNA strands of the environment in object.

Lastly, although not formally reviewed in this study, custom-formatted project dashboards are also available to every author.

8 THE PROOF OF CONCEPT CASE STUDY

In the proof-of-concept case study a large motor yacht is developed through the first phases of design and (virtual) production. Principal characteristics are derived from generic GA plans and reference to owner's requirements:

- LOA 134m
- BOA 18.5m
- D 14m
- Speed (cruising) 25kts

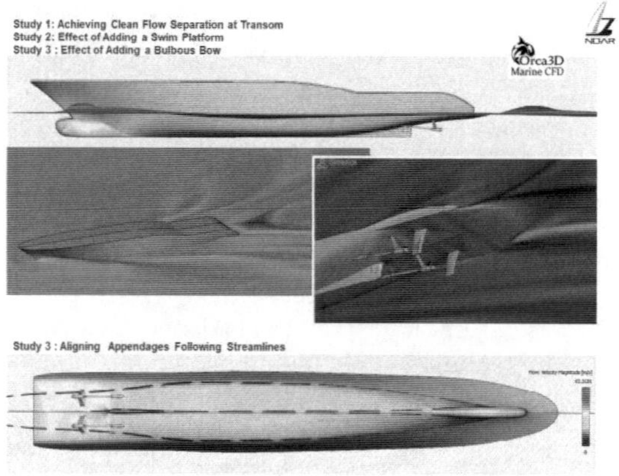

Other naval architecture values are computed accordingly and finally came to:

- LWL 125m
- DISPL 6330t
- T 5.5m
- Propellers 2
- Propeller Diameter 4m
- Power delivered at 25kts 16MW
- Gross Tonnage (approx.) 4000 t

Only the hull was addressed at this stage, it being considered that aside from its Weight and CG the modelling of the superstructure would have contributed little to the conclusion in this phase.

There is no convenient way to describe a non-linear collective process, therefore the *author-to-author* linear relations are presented in discrete format, grouped loosely in accordance with the somewhat overlapping software categorization by "purpose" described elsewhere in this document.

When considering the limited number of examples presented, the reader is invited to replace the tools named therein with her/his own which might likely yield a higher productivity environment, too, as will *any* subset of the overall environment being considered.

Furthermore, only few *author-to-author* relationship and interaction can be described, the reader is therefore asked to construe a little and implicitly consider wholly the shared, collaborative d&i space.

Similarly, a sequential representation of a non-linear process inevitably violates some priorities and sequences of events, the reader should abstract from the formality of the following "list" and construe the integrated design and production space.

Finally, crucial to the environment's high-performance, most of the tools considered support scripting and/or automated.

Unless explicitly noted otherwise, native file formats are used.

8.1 (a) "CAD" and d&i "management"

A bland distinction is made between the inevitably overlapping phases of:

- design: from concept to design-for-production reviews, classification drawings and adjustments during production and warranty service
- production: starting with feasibility, early BOMs , weld estimations, what-if studies, design-for-production reviews, etc. through delivery and warranty

Following the initial brief, the concept & feasibility design process commences:

ShipWeight: parametric weight and CG estimation, produce a full report for Naval Architect, Project Manager, etc., summaries in custom xml/text format for other authors.

Cost Fact plus *ShipWeight, ShipConstructor, ERP, etc.*: parametric cost estimation, meta-data merged with other authors'

Rhino3D, Orca3D, Orca3D CFD plus *Vray*, *Bongo*, *Penguin*, *Flamingo*, *Brazil* etc.: create fair surface model to be the parametric base for other authors. Model compartments for stability calculations. Run early CFD experiments. Produce renderings, animations of doors, deck equipment, toys, etc.

AutoCAD plus *ShipConstructor*, *Revit*, *Inventor, AdvanceSteel*, *EnterprisePlatform*, *Vray*, *3DS Studio Max*, *etc.*: read Rhino + ExpressMarine model, compose 3D GA, automatically produce the corresponding 2D drawings, BOMs, discipline 2D/3D drawings, study/review/management 3D models (full and partial). Produce renderings.

EnterprisePlatform plus *Project, Navisworks*: create and automatically maintain the project space's complete archive in accordance with planning. Use EP Operations to interacts with other scripting-compliant software, handle files generated by other authors (CAD, ERPs, DMSs, PLMs, MS-Excel, etc.) Merge with planning. Produce Unique Models.

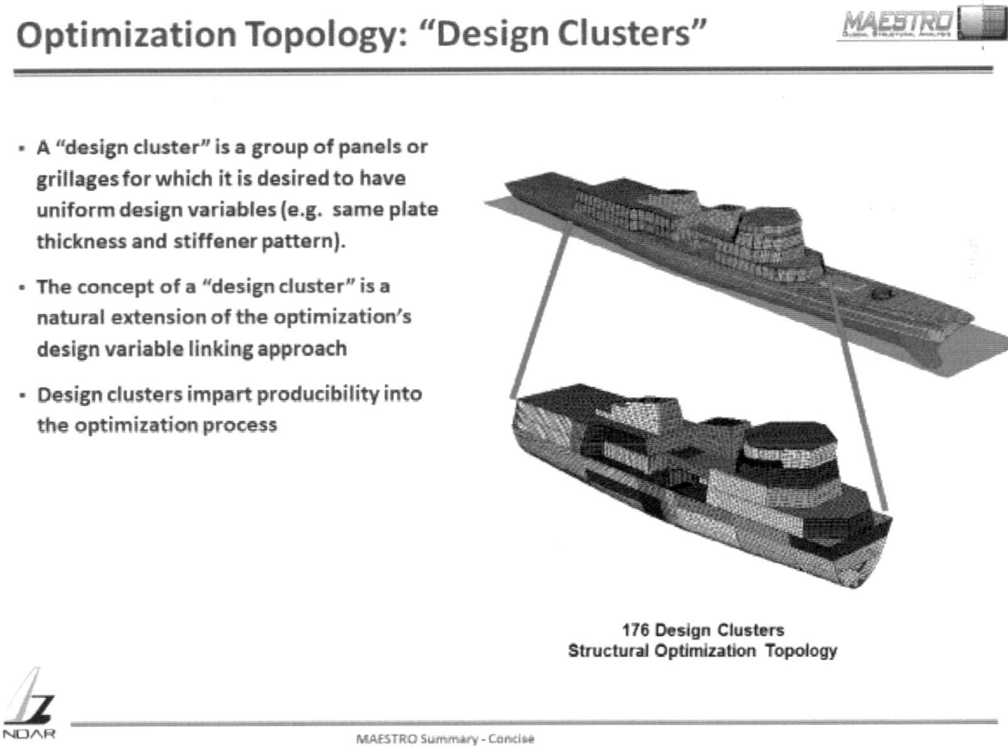

Figure 6 : MAESTRO uses *Design Clusters* to perform structural optimization for production and evaluation patches for full fatigue prediction following an integrated ship motions analysis.

Project plus *Navisworsk, ShipConstructor, other authors*: merged, interactive planning with aligned geometry and meta-data.

Rhino3D plus *ExpressMarine3D*: parametrically develop first model of non-dimensional primary and secondary structure based on hull shape and GA.

AutoFEM: read ExpressMarine model, evaluate local/medium size structural arrangements.

GHS: read Rhino hull model (with/ without compartments), read ShipWeight weight curves, define loading conditions, compute damaged and probabilistic stability, longitudinal bending moment, run seakeeping. Produce report for Naval Architect, Project Manager, text/xml summaries for other authors.

NavCad and *PropElements*: read the Rhino model (stl format), compute resistance, wave drag, propulsion train (engine, power, propeller), converge design space extents for CFD runs. Refine propeller sized in NavCad.

MAESTRO: compose global structure model, first principles scantling calculations, read ShipWeight's weight curve + GHS's loading conditions, compute time & frequency domain ship motions full fatigue screening, early optimisation for production. Read Rhino model (iges). Read ExpressMarine model, apply scantlings for first structural estimation.

Navisworks: collect 2D drawings, Rhino3D, AutoCAD, GHS, MAESTRO, ShipConstructor, ShipWeight rich 3D models for review, comparison, project management. Produce Unique Models.

Several authors: seed the VR, AR, Digital Twin, IIoT&S environments.

Data is tracked and monitored by the presence / absence / changes of files identified in a user-defined, editable event list and by tracking data and changes to it within xml files. Data will change during the occurrence of the various events, that information is relayed to the concerned author(s) by iYaldi's *iNotifika EP*, *iNotfika* and *iFileMinder* according to the user-defined list of events/recipients.

d&i is shared outside the company using any suitable shared environment, WAN, VPN, Cloud, etc. such as *DropBox*, *OneDrive*, *FTP*, etc. Email, USB sticks and similar unmanaged data transmission means are formally proscribed.

As fundamental Naval Architecture parameters converge, some authors are called upon more than others. For example:

AutoCAD plus *ShipConstructor, EnterprisePlatform, Cost Fact*: read the ExpressMarine model, create a formal structure model, advance main distributed systems model (pipe, hvac, electrical), compose general reference model, produce Class drawings and advanced documentation for detailed cost analysis, produce renderings, VR, AR models (used with mock-ups). Initiate the Digital Twin environment.

AutoCAD plus *ShipConstructor*, *Revit*, *Inventor*, *Vray*, *3DS Studio Max*, etc.: advance 3D GA, produce 2D plans, renderings, feed the VR, AR, Digital Twin platforms.

iYaldi, other shared environment software: maintain d&i alignment amongst all authors.

This leads to the detailing of the weight model and to the final structural, stability and propulsion evaluations:

ShipWeight: acquire advanced part lists with weights and CGs, continue comparison between weight budgets and weight of tracked items.

AutoFEM, GHS, MAESTRO, NavCad, etc.: carry out advanced Naval Architecture checks, final life-cycle performance estimates.

Cost Fact : increase detail level of cost prediction and analysis.

EnteprisePlatform: compose initial work packages for production.

At some point during the events mentioned above, production engineering and production will have commenced:

AutoCAD, Rhino3D, Cost Fact, Project, etc. plus *ShipConstructor*, *ExpressMarine*, *AutoFEM*, *EnterprisePlatform*, *Navisworks*: produce detailed design model and drawings, production engineering data and documents, planning, work packages, KPI trackers, specific documents to support individual departments and disciplines, etc.

AutoCAD, Rhino plus *3DS StudioMax, Vray, Brazil, Navisworks*: rendering, VR, AR.

AutoCAD, MIM plus *ShipConstructor, Navisworks, EnterprisePlatform*: Digital Twin, operational PLM model of the vessel.

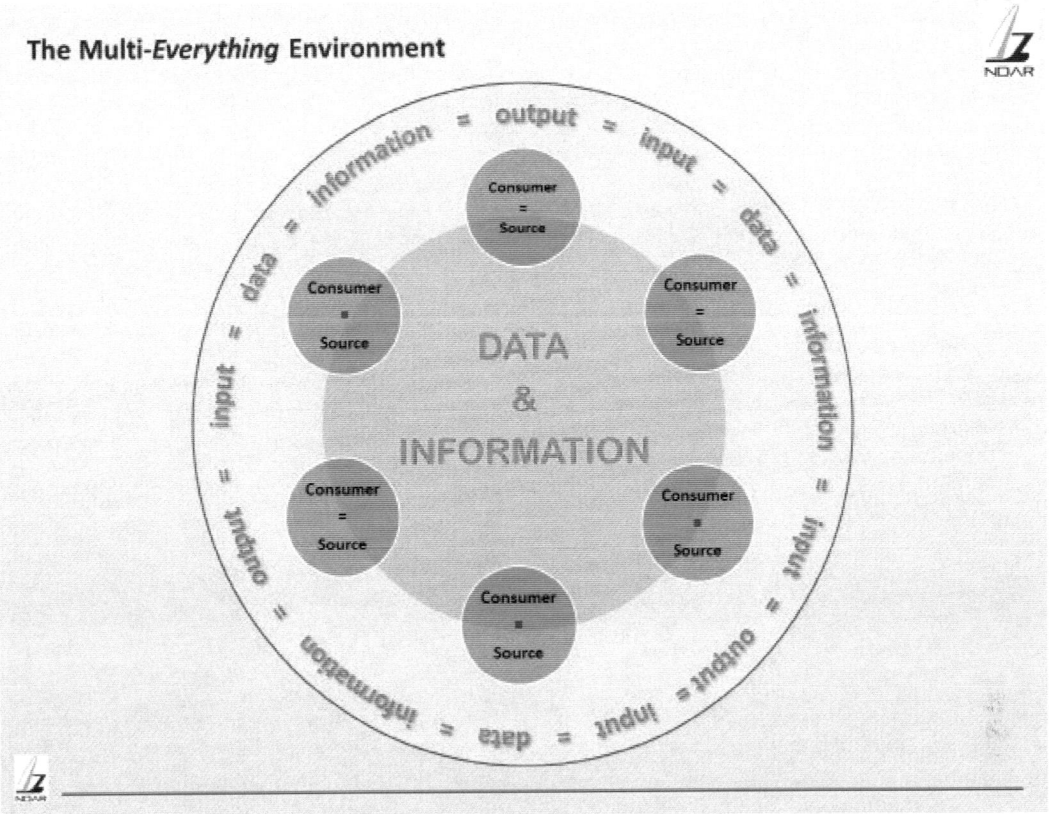

Figure 5: schematic representation of the multi-*everything* environment and d&i space.

iYaldi, other shared environment software: maintain d&i alignment amongst all authors.

8.1 (b) d&I, File & Document "Management"

Essentially, these services are underlaid by:

- EnterprisePlatform
- Project

and provided by:

- iYaldi suite
- shared environment software

8.1 (c) Rising Technology: VR, AR, Big Data, Digital Twin, IIIoT&S

Use of rising technology grows but at a slower than possible rate and generally not in a directly productive fashion. Misconceptions and limited understanding of how to make rising technologies generate ROI across the design and production environment too often leave the use of VR to the marketing and sales spheres and confine the others to the "to do" list.

Ironically, even if at the company level, big data is already used by every systems that exploits legacy d&i.

Big(ger)data is also already available *out-of-the-box* thanks to *plug & play*, consumer oriented IIIoT&S products such as Alleantia's IoT-SCADA.

The Digital Twin is a current reality, too. Software supporting rich and even intelligent 3D model with meta-data and non-CAD documents is already in use : ShipConstructor's MIM is notably Digital Twin-ready and much information can be collected from other authors, too. The Digital Twin's already in use graphical home is composed by Navisworks and one or more underlying ODBC connected databases within a shared, collaborative environment such as composed by Microsoft Azure, Dropbox, IIIoT&S, SatCom, GSM, etc.

9. CONSIDERATIONS AND PRELIMINARY FINDINGS

The preliminary findings fit well with those of other exercises characterized by both lesser and greater demands.

Although in the present *proof-of-concept* case study the full design and production cycles were obviously not completed, expectable overall performance can be estimated qualitatively and somewhat quantitatively by analogy with a comfortable degree of confidence for a design and production continuum equipped with the software used by the author, or legitimately equivalent, and a correspondingly adequate team roster and time & resource investment.

Key are found in AGILE and LEAN premises and the clear definition of scheduled internal deliverables needed to track and manage the project's progress and identify situations requiring decisions.

Another important note to make is that despite the significant d&i handling and management abilities of some software, the *out-of-the-box* environment includes some unavoidably unmanaged components that require *ad-hoc* definition, instruction and supervision.

9.1 PRE-REQUISITES

The paramount pre-requisite is the combination of management vision and will accompanied by general buy-in, overall participation and self-discipline in line with the AGILE and LEAN approaches. Other pre-requisites are:

- adoption of naming conventions for file names, file structure (layer names, chapter names, etc.), etc.
- document standards: which documents contain what, how are the contents formatted, etc.
- procedure standards: who approves what, etc.
- planning: what is to be completed, delivered, due by when, etc.
- absolute observance of the single source of truth strategy: all information exists only once, in one place, it is post-processed as required by/for the various authors
- etc.

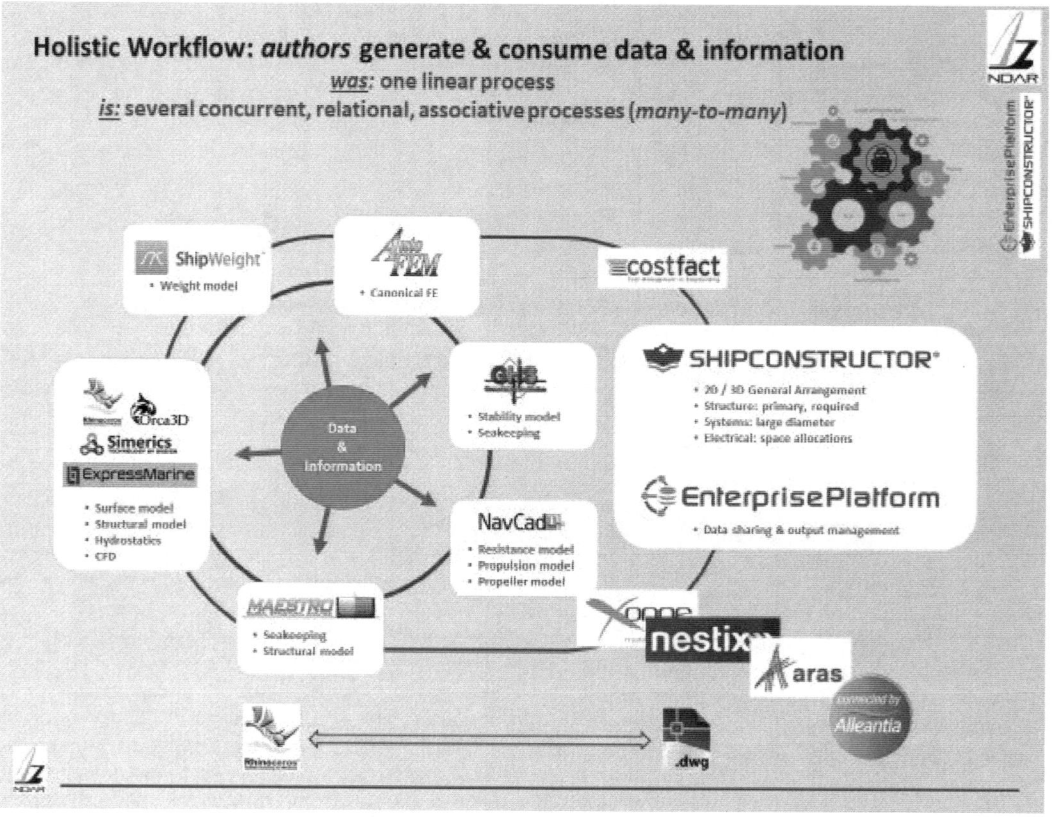

Figure 7: Very Simplified Schema of an Integrated, Multi-Author, Collaborative, Managed Environment

9.2 CONTENTS AND QUALITY

A small adjustment to the chain of responsibility will produce great value: the individual author becomes responsible for contents and quality and involves an appropriate author in case of need.

Work In Progress documents (drawings, models, Excel files, etc.), generic and custom-formatted are used for progress evaluation. Release / Versioned documents are clearly identified and stored separately, in read-only form.

9.3 EVOLUTION NOT REVOLUTION

Transforming a set of existing tools, processes and habits into an Integrated, Collaborative, Multi-Authoring, Managed Environment requires preparation and setting of short-term, attainable and immediately ROI generating goals in AGILE fashion.

A Business Process Assessment is important. In a few days, spanning enterprise-wide, it identifies objective and subjective constraints which are tackled by a following Implementation Consulting service which will firstly aim to improve existing processes, then implement new ones following LEAN and AGILE guidelines.

9.4 INVESTMENT AND TIMING

Even when using familiar tools, implementing new ways is an investment. Beyond BPA and IC, investment includes training of software users, composing and deploying an in-house dedicated team, possibly adding new software.

The plan mapped out by IC should be put into action gradually but immediately, on an on-going project, to allow measurement of immediate ROI by LEAN and AGILE metrics.

9.5 EXPECTABLE RESULTS

Expectable results obviously depend on a plethora of factors and will vary greatly from case to case and AGILE and LEAN premises cannot be stressed enough. However, experience suggests that it is reasonable to expect achievement of several goals, such as:

- overall time gain: possibly already on the first project
- error reduction: immediate, followed by:
- re-work: reduced, also depends on input, its management and changes
- quality increase: already on the first project, collective compliance with "rules" is key
- overall reduction in time and cost, increase in ROI: proportional to the above
- overall ROI: the sum of the above, estimated by some to 25%+ of total cost during year 1

Macroscopically, reduction of the cost of error and uncertainty is the quantifiable reference to be used in predicting ROI. Quantification of unjustified costs is a main BPA target.

9.6 CAVEATS

At the *out-of-the-box* level being considered in this study, the *proof-of-concept* exercise revealed the vital role played overall by:

- relational databases: an integrated model complete enough to support a large yacht project might easily exceed the million-part count, several gigabytes of database(s) and tens of thousands of files. It follows that all geometry and meta-data must be hosted in one or more relational databases in order to be further processed and managed in an automated fashion as described in this study.
- a means to manage, process, combine and make available all geometry, meta-data and ensuing documents as consumer-ready Unique Models, automatically and according to a defined schedule
- feedback loop: authors must provide their input, vital to the collaborative effort

9.7. FLEXIBILITY AND EXTENSIBILITY

The author used familiar programs to eliminate the *learning of a new tool* variable. Nonetheless, many other apt programs exist and are commonly used, too, and a non-exhaustive list is provided in this paper,

Therefore, as long as the fundamental "rules of engagement" are adhered to in AGILE and LEAN fashion, that is multiple authors communicate via native

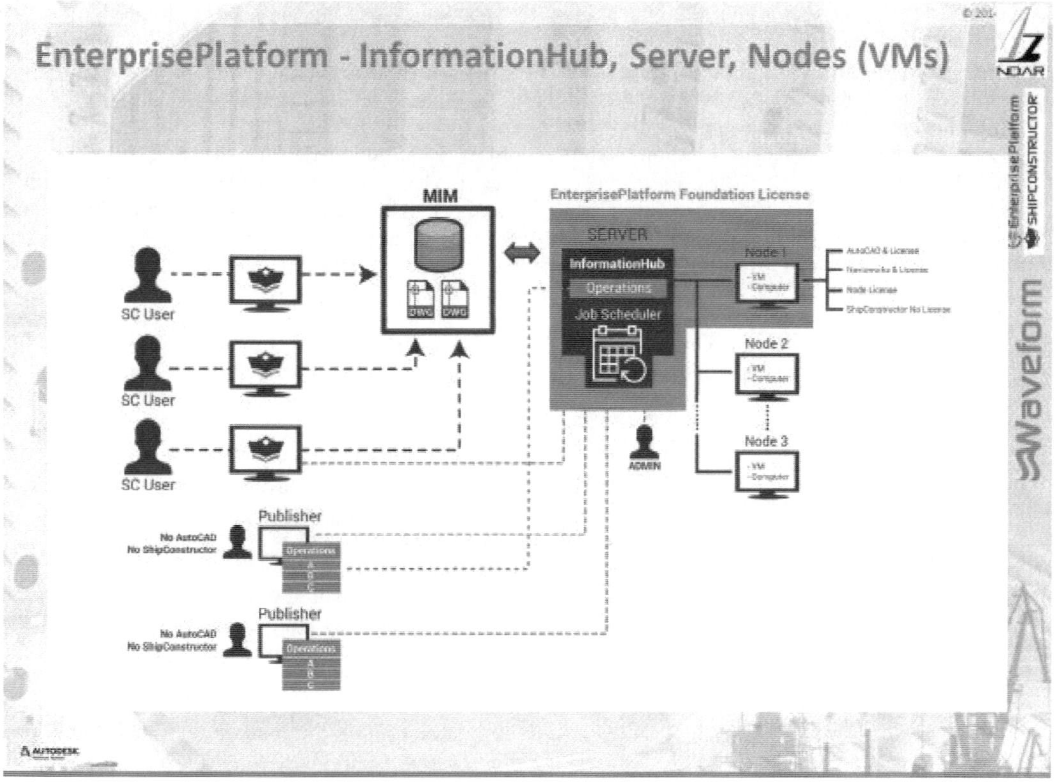

Fig. 8 – the EnterprisePlatform's functional layout

file formats, databases are used and (ideally) a d&i tracking and notification system is in place, the environment and approach presented extend easily to support many scenarios, notwithstanding that while yielding ROI at their level they do not replace a correctly implemented PLM environment.

10. CONCLUSION

The design and production of large yachts takes place in a complex, multi-author, collaborative environment that requires managed d&i sharing. Mitigation of objective and subjective constraints, introduction of adequate methods and processes, wariness of caveats and overall progressive management allow already commonly employed software to support an increase of ROI.

The approach discussed here has been experimented to varying extents in the field, with measurable benefits – for example at Royal Huisman, The Netherlands. In some instances, the exercise continued to the successful implementation of functional PLM systems – for example at Huntington Ingalls, USA. It can therefore be considered as a stepping stone, prelude to a higher-tech (PLM) environment.

It is recognised that not all the common place software tools in use today in the large yacht industry rely on relational databases and even fewer offer the means to exploit the overall d&i space. Nonetheless, some exist, and these are available to all.

Finally, other equally accessible but less common software programs exist and deserve mention within the context of the present research study, such as SSI's PLM application based on the ARAS platform. Their productive implementation, however, necessitates adequate preparation and commitment to planned investment for ROI, away from the current handicapping perception of immediate cost.

11. ACKNOWLEDGEMENTS

The authors wish to thank the innumerable peers and friends in the marine industry for their reviews, corrections and suggestions, of which only few can be mentioned here: Pedro Antunes, Volker Bertram, Lambertus Oosterveen, Siniša Relijc, Alcide Sculati, Silvano Sergiampietri, Giuseppe Sparso, Sebastian Toma, Vedran Zanic.

A special thanks goes to Denis Morais, for many, many years of inspiration, brain storming and sparring.

12. REFERENCES

[1] Danese, N., Ship CAD Systems: Past, Present and Possible Future, COMPIT, 9th International Conference on Computer and IT Applications in the Maritime Industries, TUHH Technologie, Hamburg Proceedings, 2010

[2] Oetter, R., The Role of IT in Revitalizing Brazil's Shipbuilding Industry – A Case Study, COMPIT, 9th International Conference on Computer and IT Applications in the Maritime Industries, TUHH Technologie, Hamburg, Proceedings, 2010

[3] Danese, N., CAD Centric, Multi-Source Data and Information Management, COMPIT, 10th International Conference on Computer and IT Applications in the Maritime Industries, TUHH Technologie, Berlin, Proceedings, 2011

[4] Morais, D., Waldie, M., Driving the Adoption of Cutting-Edge Technology in Shipbuilding, COMPIT, 10th International Conference on Computer and IT Applications in the Maritime Industries, TUHH Technologie, Berlin, Proceedings, 2011

[5] Holmberg, M., Hunter, S., Increasing Efficiency in the Ship Structural Design process, COMPIT, 10th International Conference on Computer and IT Applications in the Maritime Industries, TUHH Technologie, Berlin, Proceedings, 2011

[6] Ma, M., Zhao, H., Danese, N., A Method of Applying Linear Seakeeping Panel Pressure to Full Ship Structural Models, COMPIT, 13th International Conference on Computer and IT Applications in the Maritime Industries, TUHH Technologie, Liège, Proceedings, 2012

[7] Morais, D., Larkins, D., Waldie, M., An effective Approach to 3D Drawings in a 3D World, COMPIT, 13th International Conference on Computer and IT Applications in the Maritime Industries, TUHH Technologie, Liège, Proceedings, 2012

[8] Larkins, D., Morais, D., Waldie, M., Democratization of Virtual Reality in Shipbuilding, COMPIT, 13th International Conference on Computer and IT Applications in the Maritime Industries, TUHH Technologie, Cortona, Proceedings, 2013

[9] Danese, N., Aasen, R., Exploiting Weight Data to Support Engineering and Corporate Decision-Making Processes, COMPIT, 13th International Conference on Computer and IT Applications in the Maritime Industries, TUHH Technologie, Cortona, Proceedings, 2013

[10] Morais, D., Waldie, M., Larkins, D., A Scalable Approach to 3D Visualization & Review Throughout the Shipbuilding Lifecycle, ICCAS, International Conferenceon Computer Applications in Shipbuilding, Singapore, Proceedings,2013

[11] Hunter, S., Freimuth, J., Danese, N., Utilizing a Robust Fatigue Screening Process for Initial Design and Throughout the Ship Life-Cycle, COMPIT, 13th International Conference on Computer and IT Applications in the Maritime Industries, TUHH Technologie, Cortona, Proceedings, 2013

[12] Morais, D., Waldie, M., Larkins, D., Leveraging Engineering Information, COMPIT, 13th International Conference on Computer and IT Applications in the Maritime Industries, TUHH Technologie, Ulrichshusen, Proceedings,2014

[13] Aasen, R., Bjørhovde, S., Methods for Representing Conditions in a Weight Database, COMPIT, 13th International Conference on Computer and IT Applications in the Maritime Industries, TUHH Technologie, Ulrichshusen, Proceedings, 2014

[14] Kaplan, R., MacPherson, D., Buildable Propeller Geometries: Answering the Unique Challenges for Design and Manufacture, COMPIT, 13th International Conference on Computer and IT Applications in the Maritime Industries, TUHH Technologie, Ulrichshusen, Proceedings, 2014

[15] MA, M., Danese, N., Hull Girder Cross-Seciton Structural Design Using Ultimate Limit States (ULS) Based Multi-Objective Optimization, COMPIT, 13th International Conference on Computer and IT Applications in the Maritime Industries, TUHH Technologie, Ulrichshusen, Proceedings, 2014

[16] Morais, D., Waldie; M., Larkins, D., Completely Rethinking Weld Management; Leveraging 3D Models and Visualisation, COMPIT, 14th International Conference on Computer and IT Applications in the Maritime Industries, TUHH Technologie, Ulrichshusen, Proceedings, 2015

[17] Larkins, D., Waldie, M., Morais, D., Utilizing a Single 3D Product Model Throughout the Design Process, COMPIT, 14th International Conference on Computer and IT Applications in the Maritime Industries, TUHH Technologie, Berlin, Proceedings, 2015

[18] Morais, D., Waldie, M., Larkins, D., Intuitive Weld Management: Leveraging 3D Models, ICCAS, International Conference on Computer Applications in Shipbuilding, Bremen, Proceedings, 2016

[19] Morais, D., Waldie, M., Larkins, D., Innovative Methods for Leveraging a 3D Model, ICCAS, International Conference on Computer Applications in Shipbuilding, Bremen, Proceedings, 2016

[20] Morais, D., Waldie, M., Danese, N., Open-Architecture Applications: The Key to Best of Breed Solutions, COMPIT, 15th International Conference on Computer and IT Applications in the Maritime Industries, TUHH Technologie, Lecce, Proceedings, 2016

[21] Morais, D (2012-2019), Waveform Blog, SSI

[22] Morais, Waldie, Danese (2016) Ship Design, Engineering and Construction in 2030 and Beyond, HIPER, Cortona, 2016, Proceedings, Proceedings

13. AUTHORS BIOGRAPHY

Nick Danese is Managing Director of NDAR (Naval Architecture, Marine Engineering, Shipbuilding) and

Syrrkle (IIIoT&S). Involved first-hand in the design and production of large yachts since 1986, management duties subtend an active R&D engagement, specifically in the integrated information and process fields, with special attention to practical implementation in the design, production and operation environments.

Stéphane Dardel is Senior Naval Architect at NDAR, responsible for ship theory software implementation and field application. A strong background in yacht design and a specialization in the use of Rhino3D and plug-ins, AutoFEM, GHS and MAESTRO subtend close collaboration with NDAR's yacht and ship design customers.

HOME CATAMARAN: COST EFFICIENT SUPERYACHT

A.Nazarov, Albatross Marine Design, Thailand
J.Wijnants, AMASEA Yachts, Monaco

SUMMARY

The paper presents case study of 23.9m catamaran motor yacht project developed for live on board and long-range ocean cruising. The yacht is designed on the limit of legal 'small craft' range though due to catamaran scheme providing accommodation spaces comparable with larger superyachts. Approaches to styling are presented, with connection to seagoing shape and other functional requirements. Layout planning and functional zones specific for catamaran yachts are discussed, with attractive features for long range cruising, weekend boating and parties on board. General dimensioning of catamarans at early design stages is based on parametric study. Recommendations on performance prediction and related hull shape design issues are given; application of CFD methods for hull refinement is presented. Approaches and problems with stability assessment of catamarans are touched upon. Comparison of catamaran yachts with monohulls is provided, with discussion of pros and cons of each platform.

Figure 1: Exterior rendering of 23.9m power catamaran for AMASEA Yachts

1. INTRODUCTION

Centuries ago catamarans and multihulls were used by peoples of Oceania and South-East Asia to discover new horizons, providing them naturally safe but home-style transport for long ocean passages. Combining such impressive heritage and new technologies allowing to build lighter boats, only recently the advantages of multihull craft have been appreciated by international yacht industry.

Today, catamarans occupy a strong niche in the overall market as pleasure boats, rescue and special craft, diving boats, passenger ferries and water taxis. In some marinas in Australia and Asia, they already count 25% of the fleet. During last 13 years Albatross Marine Design (AMD) has developed large number of catamaran concepts and designs in 18 to 120' range, of which about 50 are launched and some of them are in series production. Most of these boats are used in Asia, also in Africa, Australia and in Europe. Such a variety of design and construction experiences associated with research became the main reason why AMD was contracted by AMASEA Yachts to design new catamaran yacht model.

In brief, the advantages of catamarans, in terms of performance, safety and functionality are [3]:

- Performance – reduced hydrodynamic resistance (in certain speed range) and thus more speed or less required power;
- Soft ride - reduced vertical accelerations on seaway;
- Safety - high initial stability, higher freeboard and flotation, duplicate propulsion units, etc.;
- Comfort – extensive space for accommodations on main deck/saloon/flybridge area.

The main disadvantages of catamarans are higher construction costs and docking fees, complications in design and construction. Catamarans should be relatively light to be hydrodynamically efficient; often larger areas cause temptations to load more... But still the biggest inconvenience with catamarans is mental prejudice of large group of boat owners against multihull craft; 'catamarans can capsize' and 'not seaworthy' are just samples of such erroneous stereotypes.

The design requirements for AMASEA catamaran (fig.1) were:

- luxury expedition catamaran yacht
- length of hull L_H=23.9m – 'RCD-max' concept
- speed 8kts economical, up to 20kts max
- range 2000+ nM
- seaworthy for trans-ocean passages
- interior layouts - customizable
- toys and tenders on board
- hull material - aluminium

Figure 2: First sketch and final rendering of 23.9m catamaran

2. ARCHITECTURE OF CATAMARANS

Design of every craft in our practice starts with preliminary sketching later taken into 3D modelling software (fig.2). Two major factors that define architecture of catamarans: they are wide and 'tall', thus generally more spacious than monohull of comparable size. Due to high initial stability, the catamarans can acquire higher/bigger superstructures starting from smaller craft length. On other side, there exists practical restriction of overall beam of catamaran $B \leq 10...12$m – for access to yacht marinas and moorings. Such limitations of beam impose general restrictions on catamarans size if applied to yachts; so with L_H>30m catamaran advantage will be diminished.

Styling issues and layouts of catamarans are also greatly affected by nature of multihull craft and catamaran layout is often designed around tunnel. Most of 'sleek-looking' catamarans turn out to have insufficient vertical tunnel clearance, or lack interior volumes.

Table 1: Design options - catamaran vs. monohull

Parameter	Catamaran	Monohull
Length of hull, L_H, m	23.9	23.9
Length of WL, L_{WL}, m	23.8	23.9
Beam of hull, B_H, m	11.0	6.7
Beam at WL, B_{WL}, m	2.4	6.5
Beam between centers of buoyancy, BCB, m	8.01	-
Canoe body draft, T_C, m	1.8	1.4
Hull depth , D, m	5.9	3.9
Light displ., M_{LCC}, t	100	75
Loaded displ., M_{LDC}, t	130	105
Power, *SHP*, h.p.	2x1920	2x750
Max speed (loaded), $v_{s.}$ kts	20	15
Passenger cabins	5	4
Passengers/crew	10/4	8/4
Total area of decks, m^2	547	269
Mass of structure, t	45	38

3. CATAMARAN DIMENSIONING

To fully realize advantages of catamaran platform, the craft should be properly dimensioned at early design stages. Very often artistic styling concepts done without parallel naval architecture study will result 'awkward monstrosity' and such craft will not possess all advantages of catamaran platform.

Parametric method described in [5] has been adapted for catamarans and applied for number of design studies. Table 1 includes some results of parametric design for catamaran, presented in this paper, with comparable monohull yacht of same hull length of 23.9m. Large difference is observed in usable living and public areas giving catamaran a priority choice as 'water home' solution.

Analysis of side windage of catamarans is presented on fig.3. On one side, we tried to avoid excessive superstructures; on other side it was desirable to create accommodations comparable with larger yachts. The dynamics of interaction of small craft with elements in severe conditions is still a 'grey area' and can't be modelled exactly. Thus, it is quite important to consider certain design ratios also while designing innovative craft. By significantly exceeding well proven proportions, the catamaran can exhibit unpredictable and dangerous properties.

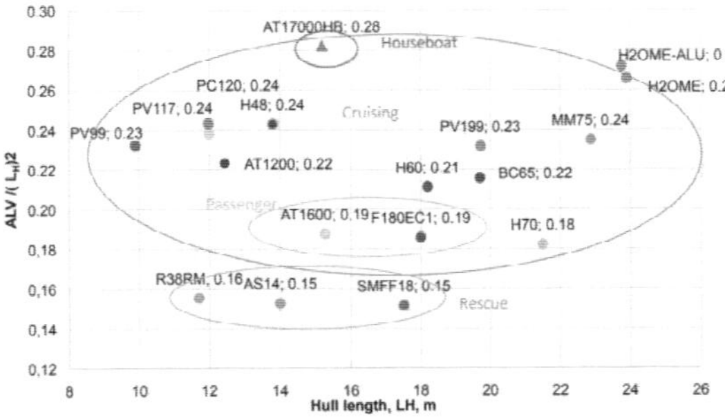

Figure 3: Analysis of relative side windage area A_{LV} of catamarans in relation to hull length L_H; for new design A_{LV}/L_H^2=0.27.

4. HULL SHAPE DESIGN

Detailed discuss of catamaran hull shape selection and parameters is presented by Nazarov in [4]. Desired speed v and 'slenderness' of hull are major factors influencing selection of hull type. Unlike monohull craft, catamarans possess enough stability due to separation of hulls and thus can feature very slender demihulls with high length to displacement ratios - relative length or slenderness ratio $l=L_{WL}/V^{1/3}$. For such slender hulls, traditional definitions of planing and displacement ranges based on volumetric Froude numbers $Fn_V=v/(gV^{1/3})^{1/2}$ are not always applicable. For very slender catamaran, $Fn_V>3$ can still present pure displacement mode where no significant hydrodynamic lift is developed. In practice, hull shapes are affected by numerous factors such as materials and construction method, preferences of designer, builder and customer, purposes of the boat, desired type of propulsion, draft limitations, etc. as well as 'design fashion'.

For AMASEA project, hull shape with 'canoe' underwater shape and stern platform (fig.4). Such shape provides excellent protection of stern gear, along with proven hydrodynamic efficiency in number of AMD's designs.

Proper dimensioning of tunnel – space between demihulls, cross deck and waterline is crucial point of catamaran design. Few tunnels shape factors should be considered:

- Vertical clearance – height t of bridge deck above water level, and longitudinal profile of tunnel;
- Transverse clearance – distance between inner sides of hulls c; or beam between centers of buoyancy B_{CB}; or beam between centerlines B_{CL};
- Tunnel shape – transverse and longitudinal sections and details of shape.

Seakeeping of catamarans is strongly limited by vertical clearance t, and performance is affected by horizontal clearance c. Considerations for vertical clearance t on displacement catamarans are [4,6]:

- Minimal vertical clearance t=(0.045…0.060)L for displacement catamarans in the middle of the tunnel, with appropriate rise at bow.
- Operational wave height is another consideration for vertical clearance t, selected to avoid slamming in anticipated conditions. From results of sea trials and computer simulations, severe slamming often starts at significant wave height $h_{1/3}$ corresponding to double of vertical clearance, i.e. $t \approx 0.5h_{1/3}$. Bow area 0.25L of cross deck should have a rise to $t \approx h_{1/3}$. This bow area can feature 'wave breaker' or can be replaced by net.
- Running trim of power catamaran can have significant effect on vertical clearance and wash of aft part of the tunnel.

For displacement catamarans transverse clearance c is greatly affected by resistance considerations, limitations for transportation/docking, accommodations, etc. Often, a reasonable compromise has to be reached sacrificing performance in favor of other properties. In general, the closer is the design speed is to the hump of resistance; the wider should be the catamaran, in order to minimize interaction of wave systems and caused extra resistance.

In terms of tunnel shape, any protrusion in the tunnel should be avoided. Unfortunately, some catamaran models possess excessive protrusions in tunnel as a trick to fit accommodations; they are excellent to live on them in marina but show poor seakeeping once at sea. Also, stern platform at centerline and its lifting mechanism can create a lot of extra drag, if not made flush with tunnel surface.

5. STABILITY

Stability is seen as main advantage of catamaran platform. However, most of existing stability requirements are oriented towards monohull craft, and stability of catamaran can be 'penalized' by improper application of norms by certifying authority. Recreational catamarans below 24m in length are covered by ISO12217-1 standard which provides specific criteria for wide craft. Though workable, design experiences shows that these requirements impose some disadvantage of high required downflooding angles on catamaran craft.

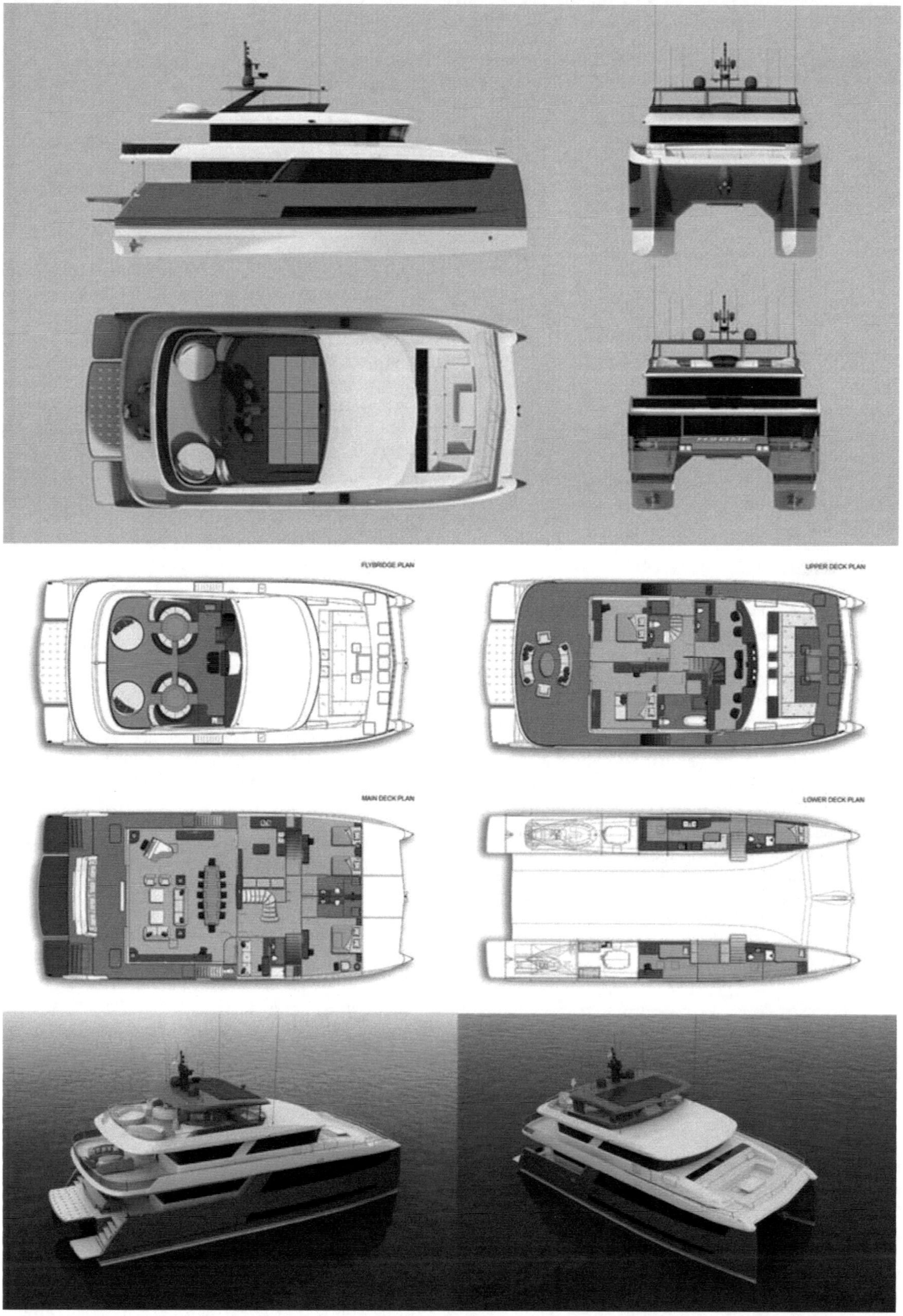

Figure 4: Design images, layouts of AMASEA 23.9m catamaran yacht

Originally, ISO stability standards for small recreational craft were of 'simplified nature' and intended to be used by average boatbuilding technician, also in form of testing. By now ISO12217-1 stability criteria are quite extensive and require analysis of 7-8 load cases, including flooding or recesses. This man hours and skills required to perform stability compliance analysis exceeds such for comparable commercial craft.

Aft garage spaces are considered floodable - they are taken as open and excluded from stability analysis geometry (fig.5).

.

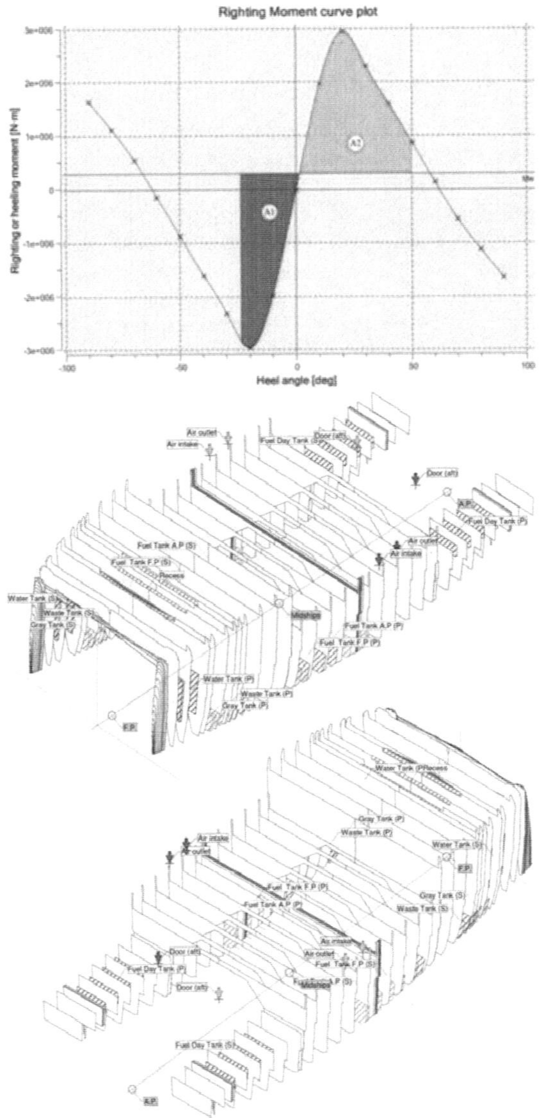

Figure 5: Stability diagram and 3D model used for stability analysis of catamaran craft, representing tanks and downflooding points.

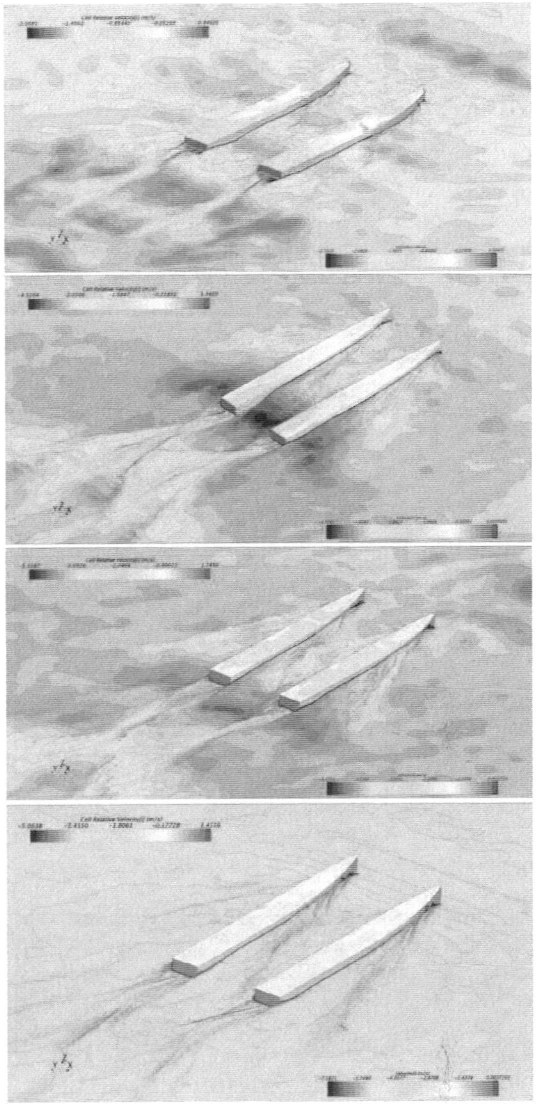

Figure 6: CFD study of catamaran rresistance and wave making, showing speeds 8, 14, 17 and 20kts

6. PERFORMANCE

Accurate resistance predictions for multihull craft are quite complicated and results can be contradictory. 'Anchoring' of computations to results of sea trials is valuable source for verification of speed-power estimates of new designs of catamarans.

For displacement catamarans additional resistance R_C can comprise (-10…40) % of total resistance; for planing catamarans about (-5…20) % of total resistance [4]. Factors, influencing specific additional resistance R_C of catamarans are following:

- Interaction of wave systems of isolated hulls, effecting mainly wave resistance;

- Increase of water flow speed between hulls, effecting mainly frictional and viscous resistance;
- Increase of wetted surface due to spray and 'tunnel riding', effecting mainly frictional resistance

The latter factor is mostly affected by details of hull shaping and tunnel proportions and is present when tunnel surface is getting wetted, mostly on high-speed catamarans. In AMD's practice, this factor is evaluated by applying frictional components from observations of flow on similar type of craft during sea trials.

For preliminary estimates, for displacement and semi-planing catamarans two methods of hull resistance prediction based on systematic series are available:

- Molland's catamaran series [1];
- Müller-Graf series [2].

The first method allows study of horizontal clearance and intended for round bilge hulls. The second method is intended for sharp chine shapes of high speed catamarans.

At advanced design stages, CFD methods are used (fig.6), allowing to optimize details of hull shape and interaction of hull with propulsors.

7. STRUCTURAL CONSIDERATIONS

Catamarans with their complicated geometry and operational profiles are subject to considerable global and local structural loads, mainly on the bottom and tunnel area. As catamarans are sensitive to weight, the concern should be given to structural optimization and weight saving. The current approach to structural design of catamarans is to consider the following loads:

- Local loads on bottom, sides, deck, cabins - estimated as for monohull boats;
- Local strength of tunnel – specific slamming loads for catamarans;
- General strength of catamaran with special attention to transverse bending, shear and torsion.
- Specific loads on details such as engine base, masts, anchoring and towing equipment, sailing rig loads, etc.

Currently, there is no acting standard for scantlings of recreational catamarans. Implementation of ISO12215-7 standard has been postponed and preliminary version is not usable for practical calculations of power catamarans. Local loads can be defined using ISO12215-5 standard (for monohull craft with hull length L_H<24m) or appropriate structural rules of classification societies. Design loads deriving from number of rules used in AMD's practice produce similar values of local loads on hulls and decks, but considerable differences exist for wet deck loads.

Longitudinal global strength is usually not an issue for catamarans with L_H<30...40m, exclusive of some craft with unusual proportions and structural schemes. Global strength of the catamaran is provided due to hull, side, tunnel and deck plating and transverse bulkheads. Standard design loads for transverse bending moment M_B, transverse torsion connecting moment M_T, transverse connecting shear force Q_T are defined as [4]:

$$M_B = k_{MB}\Delta B_{CL} n_{cg}$$
$$M_T = k_{MT}\Delta L_{WL} n_{cg}$$
$$Q_T = k_{QT}\Delta n_{cg}$$

where Δ - is weight displacement of craft; B_{CL} – distance between hulls centerlines; L_{WL} – length of waterline; n_{CG} – design value of acceleration at CG of craft; n_{CG}=2...6 depending on type of craft. The coefficients in those formulas depend on design category and standards applied, and comprise $k_{MB}=k_{QT}$=1.25...2.5; k_{MT}=0.63...1.25. For smaller catamarans or higher categories, where one of demihulls is likely to become airborne during operation, the coefficients are higher.

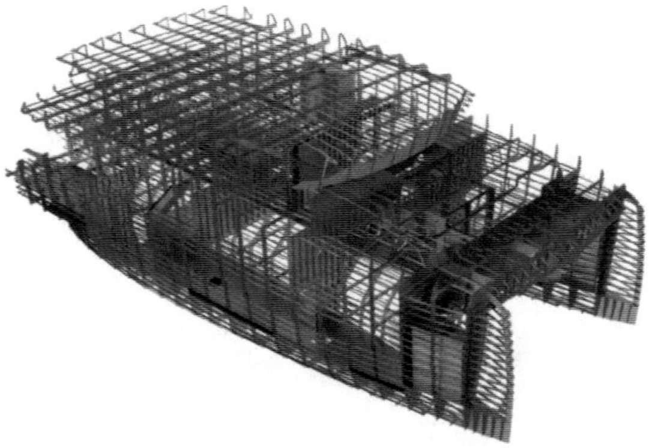

Figure 7: Structure of catamaran in aluminium alloy

Global strength analysis in AMD's practice is performed by finite elements methods (FEM) using loads specified above. However, calculations of local strength will always be the defining condition for practical engineering of catamaran structures.

8. INTERIOR LAYOUT AND STYLING

Catamaran concept offers unprecedented facilities and comfort for limited length. For AMASEA catamaran, general idea is to use flexible layouts which can be tailor-made for particular boat owner. This is possible due to 'square' shapes of catamaran volumes.

Figure 7: Options of interior styling for master cabin (above) and saloon (below)

Another concept is that owner can bring his interior designer into project, allowing custom interior styling and creating unique yacht. Some possible options are shown on fig.4 and fig.8.

Lower deck is used for crew and personnel, with their own essential facilities such as galley. Main deck carries spacious saloon, office and guest cabins, and also cooking/serving facilities. Upper deck accommodates pilothouse, and one or two master cabins. Both upper deck and flybridge offer plenty of outdoor spaces smartly planned with sitting areas. The catamaran caries toys on board - there are two garage spaces.

9. CONCLUSIONS

Catamarans have proved to be fast and reliable craft with plenty of operational advantages for different applications – pleasure, special and commercial craft. Development of superyachts on catamaran platform allows some obvious benefits in range of lengths L_H=24…30m or even larger. Yacht building industry, perspective owners and marina operators should take a closer look at superyacht catamarans and catamaran-friendly facilities.

Cost effective AMASEA catamaran with length just below 24m is ‘small recreational craft’ in terms of legislation with simplified registration and manning requirements. However, this offers superyacht facilities in terms of space and comfort. Such boat would make a statement in any marina due to its visual impression of size.

10. REFERENCES

1. Molland A.F., Wellicome J.F. Couser P.R. Resistance experiments on a systematic series of high speed displacement catamaran forms: variation of length-displacement ratio and breadth-draught ratio. Transactions of the Royal Institution of Naval Architects, Vol. 138, 1996.
2. Müller-Graf B. Widerstand und hydrodynamische Eigenschaften schneller Knickspant-Katamarane der VWS Series '89 (Resistance and Hydrodynamic Characteristics of the VWS Fast Hard Chine Catamaran Series '89)// 20th Symp.Yachtentwurf und Yachtbau, Hamburg 1999, Germany.
3. Nazarov A., Jabtanom P., Phormtan P., Leeprasert A., Suebyiw P. Catamarans as Superyachts// Design and Construction of Super & Mega Yachts, Genoa, Italy - 8-9 May 2013.- P.183-192.
4. Nazarov A. Small Catamarans: Design Approaches and Case Studies//International Journal of Small Craft Technology, RINA, Vol.157, Part B1 2015.
5. Nazarov A. On Application of Parametric Method for Design of Planing Craft //3rd Chesapeake Power Boat Symposium, USA – June 2012.
6. Yun L., Bliault A., Rong, H.Z. High Speed Catamarans and Multihulls: Technology, Performance and Applications. Sringer, 2019.

9. AUTHORS BIOGRAPHY

Albert Nazarov holds the degree in Naval Architecture from Sevastopol National Technical University (1996) and Ph.D. in Ship Dynamics from Odessa National Maritime Academy (2004), also diploma in superyacht construction management from Maritime Training Academy (2015). His current position is managing director at Albatross Marine Design with experience in construction, design and research of small and high-speed craft.

Jack Wijnants holds a Master in Electronics, Telecommunication and ICT Engineering Technology from the High School in Antwerp (1987). He is managing director of MSAR Ltd in Malta since 2010, constructor of Search and Rescue Vessels and other safety equipment. In 2018 he launched the Amasea Yacht brand in Monaco to commercialise the three deck H_2Ome (waterhome) catamaran.

HYDROFOIL CONFIGURATIONS FOR SAILING SUPERYACHTS: HYDRODYNAMICS, STABILITY AND PERFORMANCE

J-B R G Souppez, Solent University, UK.
J M M-A Dewavrin, Gurit, UK.
F Gohier, Bénéteau Group / ENSCBP University, France.
G Borba Labi, Principal Power, Portugal.

SUMMARY

Hydrofoil-assisted racing monohulls have undergone significant development phases in the past decade, yet very little scientific data has reached the public domain: an increasingly critical issue as the superyacht industry is now looking at the implementation of foils onto leisure vessels. Consequently, three contemporary configurations, namely a Dynamic Stability System, a Dali-Moustache and a Chistera have been towing tank tested to present the first complete characterisation of the hydrodynamic efficiency, quantification of the added dynamic stability and eventually the resulting impact on sailing performance. Furthermore, the considerations inherent to the design and installation of hydrofoils onto superyachts will be detailed. Building on extensive experimental work, this paper provides a comprehensive assessment of current design options with both technical and practical guidelines and recommendations to improve performance.

NOMENCLATURE

$1 + k$	Form factor (-).
A	Planform area (m^2).
B_{OA}	Beam overall (m).
B_{WL}	Beam on waterline (m).
$\bar{c}$	Mean chord (m).
C_T	Total resistance coefficient (-).
D_{WL}	Design waterline (m).
Fn	Froude number (-).
F_H	Side force (N).
L_{OA}	Length overall (m).
L_{WL}	Length on waterline (m).
R_I	Induced drag (N).
R_T	Total resistance (N).
s	Span (m).
t	Temperature (°C).
T_C	Canoe body draft (m).
T_{EFF}	Effective draft (m).
T_K	Keel draft (m).
U	Uncertainty (-).
V	Velocity (m/s)
WSA	Wetted surface area (m^2).
α	Sweep angle (°).
θ	Heel angle (°).
λ	Leeway angle (°).
ρ	Density (kg/m^3).
AoA	Angle of Attack.
CNC	Computer Numerically Controlled.
DSS	Dynamic Stability System.
DSYHS	Delft Systematic Yacht Hull Series.
FP	Forward Perpendicular.
IRC	International Rating Certificate.
ITTC	International Towing Tank Conference.
LCG	Longitudinal Centre of Gravity.
NACA	National Advisory Committee for Aeronautics.
ORC	Offshore Racing Congress.
VPP	Velocity Prediction Program.

1. INTRODUCTION

The implementation of hydrofoils on leisure vessel was first featured in 1898 on powerboats, before being employed on a sailing catamaran in 1938 under the leadership of the National Advisory Committee for Aeronautics. Then, circa 1954/1955, foiling monohulls emerged, with the Baker Manufacturing Company building various size dinghies. Eventually, the 1960s saw their use in offshore racing. Nevertheless, despite their historical use, the last decade sparked an unprecedented regain of interest, with hydrofoiling yachts featured in several forms in the most competitive and prestigious sailing events, from the America's Cup to the Vendée Globe.

While significant numerical and experimental work has been conducted by the design and race teams, hardly any technical data has been made publicly available. Consequently, this paper aims to remedy this absence of open source information by providing results for different foil-assisted monohull configurations, whilst also tackling performance prediction and design consideration for their implementation on superyachts.

Firstly, the previous work, aims and objectives, and the foils will be introduced, followed by a description of the experimental setup, as well as the design and manufacturing considerations for the model and three hydrofoils: a Dynamic Stability System, a Dali-Moustache and a Chistera. Then, the towing tank results will be presented in different conditions, representative of upwind and downwind sailing, eventually discussing the hydrodynamic efficiency, added dynamic stability provided, and the overall effect on the performance of the vessel. The advantages and drawbacks of each option will be outlined, finally concluding on practical design considerations and recommendations.

2. BACKGROUND

2.1 PREVIOUS WORK

For offshore racing monohulls, the literature has primarily been focussed on the long-established use of straight asymmetric daggerboards, as summarised by Campbell *et al.* (2014). On the other hand, the design of hydrofoils for flying dinghies, such as the International Moth class, have been extensively investigated (Beaver & Zselczky, 2009). Furthermore, new research emerged in the last few years, targeted at flying catamarans and the optimisation of flexible foils (Sacher *et al.*, 2017) and issues associated with ventilation (Binns *et al.*, 2017), all heavily influenced by the developments in the America's Cup. The literature, however, does not tackle foil-assisted monohulls.

The past couple of years also saw the first large scale production of an offshore racing vessel with hydrofoils, namely the Figaro Bénéteau 3, and more recently the first superyacht fitted with a Dynamic Stability System (DSS), namely the Baltic 142. Moreover, 2018 marked the addition of foil measurements as part of the International Rating Certificate (IRC) racing rule, reflecting contemporary practice in racing craft design. This shows the strong interest of yacht and superyacht designers for foiling technology, and the necessity for published data relative to their efficiency, stability and overall effect on performance.

2.2 AIMS AND OBJECTIVES

The experimental investigation into foil-assisted monohulls aims to quantify the hydrodynamic efficiency and ascertain the added dynamic righting moment provided, to ultimately predict the velocity. Three main contemporary designs will be tackled, namely the DSS, the Dali-Moustache and the Chistera foils.

2.2 (a) Dynamic Stability System

The DSS is a retractable transverse foil deployed to leeward, the intention being to increase the righting moment, but also to reduce the pitching moment, allowing a more comfortable sailing. Unlike the Chistera and Dali-Moustache foils, the DSS only provides vertical lift due to its solely horizontal planform.

2.2 (b) Dali-Moustache

Based on the IMOCA racing yacht design, the Dali-Moustache is a V-shaped foiling daggerboard, intended to improve stability, while contributing to both the side force and vertical lift, the latter reducing the effective displacement of the vessel. The other advantage of the foil is the decrease in the pitch angle of the boat, improving the longitudinal stability and sea-kindliness (*i.e.* damping the pitch motion).

2.2 (c) Chistera

Finally, the Chistera foil is based on the Figaro Bénéteau 3 one-design class. In contrast with the Dali-Moustache, the Chistera has an inward-facing V-shape, that also provides both vertical lift and horizontal side force, together with additional righting moment.

3. EXPERIMENTAL TESTING

3.1 MODEL

The tank testing of the different configurations has been performed on a purposely designed hull (Dewavrin, 2018), first towed bare, before the keel and bulb were added; finally, each foil was evaluated. The main dimensions for the 1:10 scale model, representative of a 50ft sailing yacht then use to extrapolate the findings onto superyachts, are presented in Table 1

Hull Particulars	
Length overall - L_{OA}	1.52 m
Length on waterline - L_{WL}	1.43 m
Beam overall - B_{OA}	0.47 m
Beam on waterline - B_{WL}	0.34 m
Canoe body draft - T_{C}	0.06 m
Keel draft - T_{K}	0.36 m
Wetted surface area - WSA_{H}	0.39 m^2
Keel Particulars	
Span - s_{K}	0.266 m
Mean chord - $\bar{c}_{\mathrm{K}}$	0.068 m
Planform area - A_{K}	0.018 m^2
Wetted surface area - WSA_{K}	0.037 m^2
Section	NACA 64-012
Swept back angle - α	3°
Leading edge distance aft of FP	0.636 m
Bulb Particular	
Chord - $\bar{c}_{\mathrm{B}}$	0.270 m
Wetted surface area - WSA_{B}	0.023 m^2
Horizontal section	NACA 65-017
Vertical section	NACA 65-012

Table 1: Tank testing model dimensions.

General modelling and scaling laws are driven by Froude's similitude theory. Equality in Froude number between model and full-scale will ensure that gravity forces are correctly scaled. However, this implies that the vessel and appendages will operate at a too small Reynolds number, thus not replicating the full-scale laminar to turbulent transition. As a result, transition will artificially be triggered using sandpaper strips, in accordance with the International Towing Tank Conference (ITTC) procedures (ITTC, 2017).

3.2 HYDROFOILS DESIGN AND LOCATION

The general dimensions and locations of the hydrofoils were based on a parametric study of the existing vessels they are featured on. The cross-sectional shape is a critical design consideration as it directly affects the lift and drag characteristics. For consistency, and in order to compare the hydrodynamic results, the same section was employed for each foil, namely the NACA 63-412. This is commonly used for small craft, such as the International Moth (Beaver & Zselczky, 2009) and was chosen due to its high lift to drag ratio (Abbott & Doenhoff, 1959) and the relative ease of manufacturing.

Table 2 presents the main dimensions for the three foils and their leading-edge location, longitudinally aft from the forward perpendicular (FP) and vertically upwards from the design waterline (D_{WL}). Note that the spans given are for the entire foil, not accounting for its actual immersion at a given heel angle.

Dynamic Stability System	
Span - s_{DSS}	0.232 m
Mean chord - $\bar{c}_{\text{DSS}}$	0.070 m
Planform area - A_{DSS}	0.016 m²
Wetted surface area - WSA_{DSS}	0.034 m²
Leading edge distance aft of FP	0.742 m
Leading edge height above D_{WL}	-0.016 m
Dali-Moustache	
Span - s_{DM}	0.368 m
Mean chord - $\bar{c}_{\text{DM}}$	0.058 m
Planform area - A_{DM}	0.021 m²
Wetted surface area - WSA_{DM}	0.045 m²
Leading edge distance aft of FP	0.488 m
Leading edge height above D_{WL}	-0.016 m
Chistera	
Span - s_{C}	0.364 m
Mean chord - $\bar{c}_{\text{C}}$	0.056 m
Planform area - A_{C}	0.020 m²
Wetted surface area - WSA_{C}	0.043 m²
Leading edge distance aft of FP	0.488 m
Leading edge height above D_{WL}	0.142 m

Table 2: Model foil dimensions.

The positions of each foil along the hull can be visualised in Figure 1 (a), with underwater views of the DSS, Dali-Moustache and Chistera respectively shown in Figures 1 (b), 1 (c) and 1 (d) respectively. Note the forward position of the Dali-Moustache: unlike the racing yachts, it is located further forward to fit within the overall beam when retracted, and importance consideration for leisure vessel, further discussed in Section 7.3.

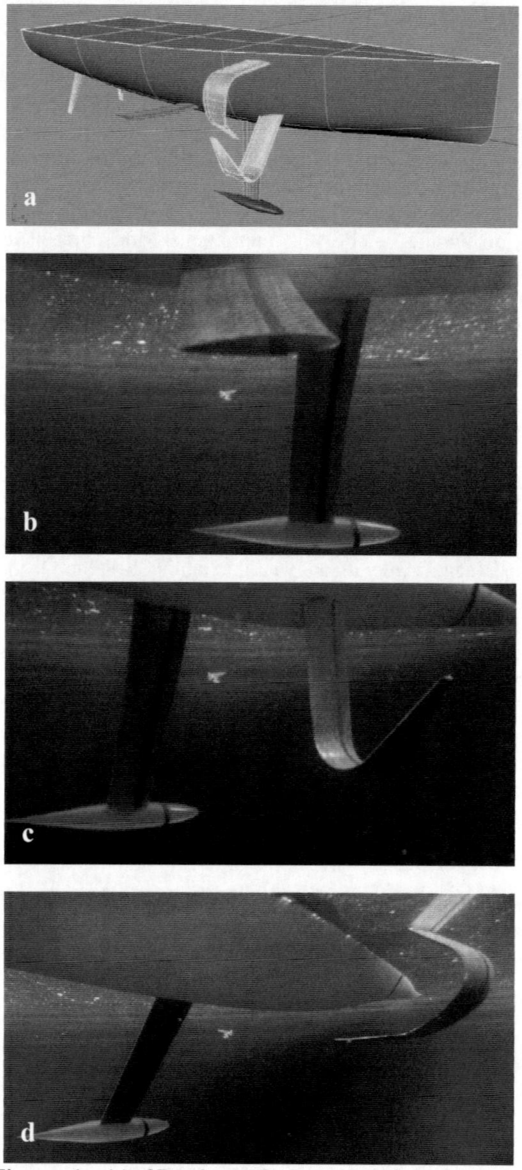

Figure 1: (a) 3D view of the appendages on the designed model. Underwater view of (b) the Dynamic Stability System, (c) the Dali-Moustache and (d) the Chistera foil.

3.3 MANUFACTURING

The hull shape was CNC cut on a 5-axis milling machine out of 32 kg/m³ polystyrene. The hull was hand laminated with two layers of E-glass woven roving having a total combined dry weight of 300 g/m² and epoxy resin. Then, it was sanded to a smooth finish, equivalent to that achieved by 400 grit wet and dry sandpaper, as per the recommended ITTC procedure (ITTC, 2017). Geometric tolerances were well within the required allowable +/- 1 mm for the overall length, breadth and depth (ITTC, 2017).

The keel was constructed out of thin laser-cut plywood, then laminated and faired. One outer layer of epoxy resin was applied for coating.

The model keel bulb and hydrofoils were manufactured out of ABS resin using stereolithography on a ProJet 3600 Max 3D printer. This was required to achieve the necessary +/- 0.2 mm tolerance on such complex 3D geometries (ITTC, 2017). Moreover, their location was accurately ascertained to respect the permitted 0.5 mm variation in position (ITTC, 2017). To strengthen the foils and ensure no deformation under dynamic loading, a layer of high modulus 200 g/m^2 twill carbon fibre and epoxy resin was applied and vacuumed consolidated at 1 atm.

Finally, all components were fitted with a 5 mm wide sandpaper strip located to replicate the full-size flow regime, as the model hull and foils would be operating at a much lower Reynolds number in the towing tank. Indeed, while the Reynolds effects on hydrofoils are not well-understood and consequently there is no current full-size correction for a smaller geometry being tested, the best practice across fields of fluid dynamics is to ensure that transition is replicated at model-scale where expected at full-scale. The use of studs or sandpaper strips to artificially trigger transition is, therefore, deemed suitable (Jackson & Hawkins, 1998), and is recommended by the ITTC (ITTC, 2017).

The locations of the rough strips were established based on the ITTC recommended Reynolds number as a function of the model/appendages length and Froude number (ITTC, 2017).

3.4 EXPERIMENTAL SETUP

The experiments were performed following the ITTC Recommended Procedures and Guidelines for Resistance Test (ITTC 2014) and were undertaken in the Hydrodynamic Test Centre at Solent University. The main characteristics of the towing tank utilized are presented in Figure 2.

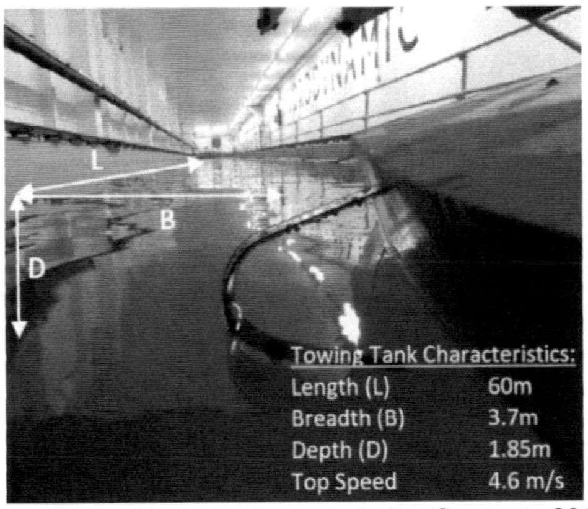

Figure 2: Towing tank characteristics (Souppez, 2018).

For the characterisation of the hydrodynamic efficiency of each foil, the runs were performed for a defined speed, at a constrained heel and yaw angle, with the vessel free to heave and trim. Conversely, to quantify stability, the model could heel freely, as later described in Section 5. The drag, side force, heave and trim (or heel for the stability investigation) were measured with a precision of five decimal places, and the data sampled at 1000 Hz over a minimum of 6 seconds, or longer at the lowest speeds where a greater data acquisition window was available.

The installation of the model on the towing carriage and the measurement devices are depicted in Figure 3. The drag, side force and trim are measured by potentiometers (P), while the heave is quantified thanks to a linear variable displacement transducer (LVDT).

Figure 3: Model installed on the towing carriage (Dewavrin & Souppez, 2018).

3.5 TEST MATRIX

The test matrix was defined after running a standard Velocity Prediction Program (VPP), where the hydrodynamic model was based on the Delft Systematic Yacht Hull Series (DSYHS) (Keuning & Katgert, 2008).

The intention was to establish a relevant set of testing parameters representative of upwind sailing on the one hand (low speed, high heel, high leeway), and downwind sailing on the other (high speed, low heel, low leeway), with also higher Froude numbers to be more in line with the performance of racing yachts (0.35 to 0.70).

Additional tests were undertaken in the first place to establish the form factor, 1+k, based on the Prohaska method suggested in the ITTC procedure (ITTC, 2014). Moreover, a preliminary study investigated the best Angle of Attack (AoA) for each geometry. In this instance, the AoA is defined as the angle between the chord line of the foil at its root and the design waterline.

Once acquired, the model scale data was scaled up to full-size (ITTC, 2011). However, prior to comparing the results for each configuration, an uncertainty analysis was performed to ensure the reliability of the data collected.

3.6 UNCERTAINTY ANALYSIS

Based on the ITTC recommended procedures and guidelines for Type A uncertainty analysis (ITTC, 2014), the experimental precision could be quantified. The parameters under consideration are the wetted surface area (WSA), speed (V), water density (ρ), total resistance (R_T) and associated coefficient (C_T). The uncertainty U, of each parameter i, and inherent components j, is labelled $U_{(i,j)}$. An example of broken-down uncertainty analysis for a resistance test undertaken at 2.25 m/s (Fn = 0.60) is shown in Table 3.

Wetted Surface Area – *WSA* (m²)	**0.453**
Model uncertainty - $U_{WSA,MOD}$	0.781%
Displacement uncertainty - $U_{WSA,BAL}$	0.025%
Wetted surface area uncertainty - U_{WSA}	0.782%
Velocity – *V* (m/s)	**2.322m/s**
Calibration uncertainty - $U_{V,CAL}$	0.002%
Data acquisition uncertainty - $U_{V,DAQ}$	0.002%
Velocity uncertainty - U_V	0.003%
Density – *ρ* (kg/m³)	**998.403**
Temperature - t	19°C
Temperature error - E_t	1.316%
Density uncertainty - U_ρ	0.010%
Total Resistance - R_T (N)	**11.049**
Calibration uncertainty - $U_{R_T,CAL}$	0.002%
Fitting uncertainty - $U_{R_T,FIT}$	1.288%
Data acquisition uncertainty - $U_{R_T,DAQ}$	4.937%
Misalignment uncertainty - $U_{R_T,MIS}$	0.934%
Resistance uncertainty - U_{R_T}	5.186%
Total Resistance Coefficient - C_T	**0.024**
Resistance coefficient uncertainty - U_{C_T}	6.245%

Table 3: Example of uncertainty analysis.

4. HYDRODYNAMICS

4.1 ANGLE OF ATTACK INVESTIGATION

Early tests were conducted to investigate the impact of the AoA of the foils. By design, they can be given a pre-set angle; many racing yachts are also typically able to adjust foils by up to +/- 7°; thus, a smaller study investigating the performance at a range of AoA was devised (Kitching, 2018).

The DSS was set at 0°, 4° and 8° AoA, while the Dali-Moustache and Chistera were tested with 0°, 8° and 16° AoA. It is important to mention that the angles defined here are at the root of the foil, the portion that would be controlled on the yacht. In the case of the Dali-Moustache and Chistera, these do not reflect the actual angle adopted by the hydrofoils, which is smaller due to the curvature and twist. The aim is to assess the optimum AoA, to then perform all the tests in their respective ideal condition, thus comparing the best possible performance for each configuration.

The investigation revealed that, when using a DSS, while a larger AoA resulted in an increase in heave and a reduction in displacement, this came at a cost in terms of resistance. Overall, a DSS with no AoA appeared to be the best solution. This is consistent with the properties of the NACA 63-412 foil that exhibits the highest lift to drag ratio at 4° for the tested Reynolds number. Despite the foil having no initial AoA, the vessel trim, ranging from 1° at low speeds to 5° at higher speeds, implies the section will naturally operate close to its most efficient AoA. It could, however, be deemed appropriate to offer some degree of control in order to alter the angle at low speed, and reduce it for the higher downwind speeds, while retaining the optimum operating angle.

For the Dali-Moustache, an increase in AoA did contribute to an increment in heave, resulting in a lower resistance. This was achieved for an AoA of 8° in upwind conditions (θ=20°, λ=2°+) and 16° downwind (θ=10°, λ=0°), with however a decrease in side force. Variations in the AoA, therefore, alter the contribution of the lift that goes towards the side force or heave. This is particularly interesting as these foils are fitted on canting-keel yachts. Upwind, the fully canted keel will provide vertical lift but less side force; which the Dali-Moustache could easily make up for.

Finally, the Chistera exhibited a better side force and heave with an angle of 16°. The impact on resistance was nevertheless negligible, thus suggesting better sailing performance will be achieved with a higher AoA.

As a result, it can be stated that for best performance, the DSS should be operated at the lowest AoA possible, while the Chistera is more efficient at a higher AoA, ensuring stall is not reached. As for the Dali-Moustache foil, variations in AoA allow to either boost the side force and reduce the heave, sensible for upwind, or raise the vertical lift at the expense of the side force, a suitable option for downwind. Consequently, the rest of the study was conducted with the most efficient AoA for each foil configuration and sailing condition.

4.2 INDUCED DRAG FACTOR

The performance of appendages can be quantified by plotting the induced drag factor, *i.e.* the side force squared versus the total resistance. For the results to be meaningful, they must be compared to the typically required upwind side force.

In this instance, the 'upwind sailing' line corresponds to the vessel operating in 16 knots of true wind (*i.e.* the upper end of Beaufort 4, after which the boat would be expected to reef), at a true wind angle of 35°. The results in typical upwind sailing conditions are presented in Figure 4 (data point at $\lambda = 6°$ not shown for the Dali-Moustache).

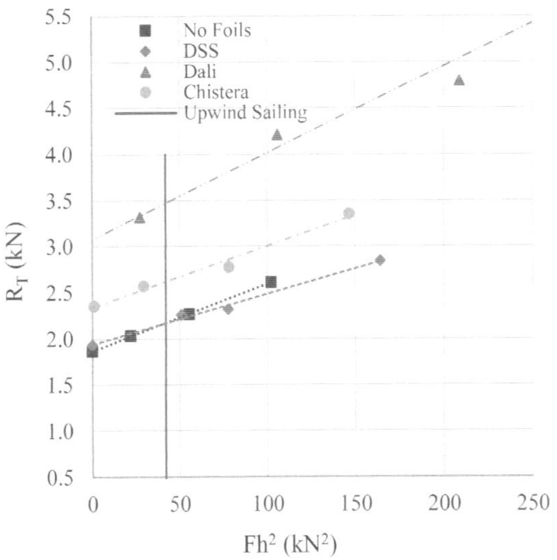

Figure 4: Induced drag factor for a typical upwind condition, Fn = 0.35 and θ = 20°; data points at λ = 0°, 2°, 4° and 6°.

Firstly, it is interesting to notice that the Dali-Moustache is the only one able to provide significant side force with no leeway. While this is no surprise for the keel alone or DSS, it could have been expected of the Chistera to be able to generate more side force thanks to its asymmetric profile without any leeway. The present results, however, demonstrate it is not the case.

Regarding the contribution of each foil to the overall side force upwind ($\theta = 20°$, $\lambda = 4°$), the Chistera provides 15% and the Dali-Moustache 45%. Those values are consistent from Froude numbers for 0.35 to 0.50.

The best performance in terms of generating side force for minimum drag is achieved by both the keel alone first, and then the DSS. However, looking at the side force that would be required to sail upwind, the keel only is superior in that portion where the realistic operation of the vessel would occur. Furthermore, this is assuming the keel only contributes to the side force, thus neglecting the asymmetry of the waterplane area, the rudder (if weather-helm is achieved), and foil (if fitted).

Under the limitations presently considered, the configuration without any foils appears more hydrodynamically efficient. Nevertheless, despite creating more resistance, the Dali-Moustache and the Chistera would contribute to reducing the leeway angle; this could permit the vessel to sail a shorter distance on an upwind course.

4.3 EFFECTIVE DRAFT

The hydrodynamic performance of yacht appendages is quantified using the effective draft, T_{EFF}, derived from the theory of induced drag on a lifting surface, mathematically:

$$T_{\text{EFF}} = \sqrt{\frac{F_{\text{H}}^2}{\pi \rho V^2 R_{\text{I}}}}$$

Where:

- T_{EFF} Effective draft (m).
- F_{H} Side force (N).
- ρ Density (kg/m³).
- V Velocity (m/s).
- R_{I} Induced drag (N).

It can be noted that the ratio $F_{\text{H}}^2/R_{\text{I}}$ is, in fact, the reciprocal of the induced drag factor slope. The DSS having the lowest slope, it naturally yields the highest effective draft, as presented in Figure 5.

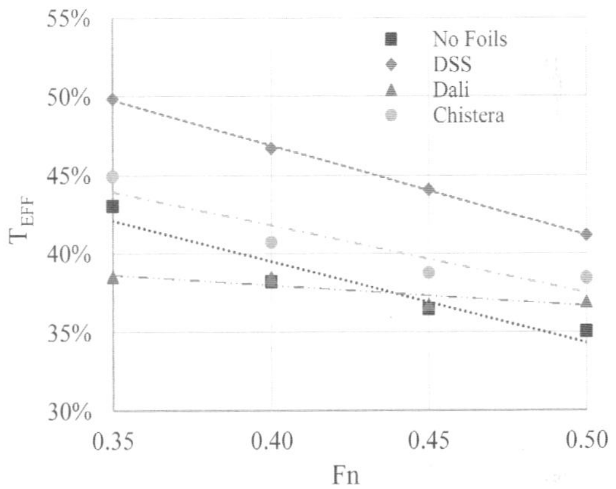

Figure 5: Effective draft at θ = 20°.

Those results should, however, be moderated with the previously identified fact that, within the normal sailing operation, the best configuration is achieved without foils. It would, therefore, be recommended that the best design option is assessed solely on the induced drag factor and in relationship with the expected side force to be provided in upwind conditions, as in this case the use of the effective draft has been proven to be misleading.

4.4 HEAVE

So far, the data analysis has been focused on the total drag and side force, critical upwind, but not accounting for the vertical lift generated by the foils. The measured heave, in both upwind and downwind conditions, is presented in Figures 6 (a) and 6 (b) respectively, where 0 heave corresponds to the static heave of the vessel.

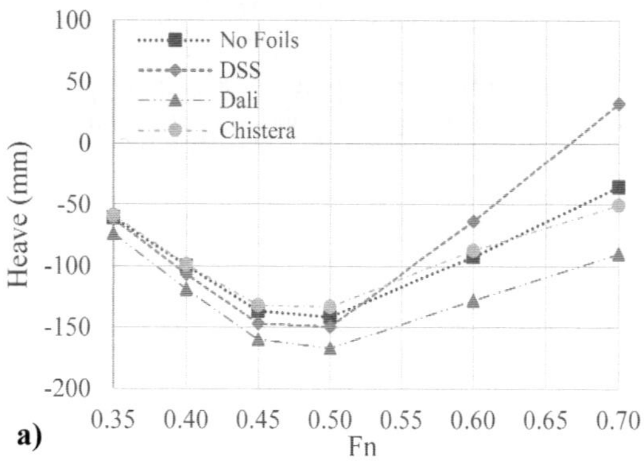

a)

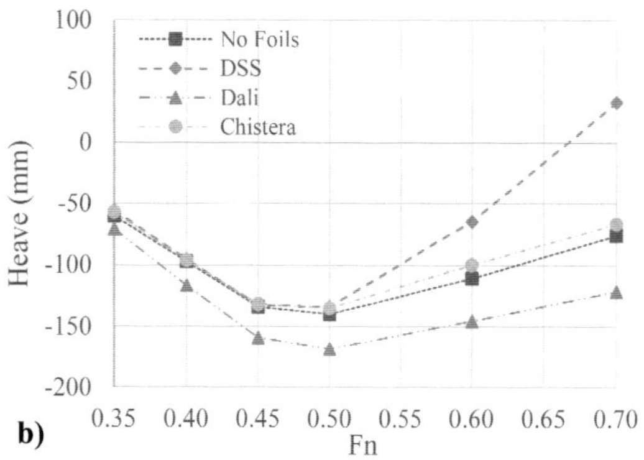

b)

Figure 6: Heave for (a) upwind (θ=20°, λ=4°) and (b) downwind condition (θ=10°, λ=0°).

The DSS, that primarily generates lift upwards, appears to be the best at reducing the effective displacement of the vessel. Moreover, due to its presence closer to the longitudinal centre of gravity (LCG), a greater portion of the lift contributes to reducing the displacement, although negligible for typical cruising Froude numbers. Conversely, the Dali-Moustache and Chistera produce a higher trim, since they are located further forward and thus the lift induces a higher pitch moment. On the other hand, the Dali-Moustache, which proved to generate the most side force (albeit with a drag penalty) did not appear to significantly lift the vessel out of the water and was recorded to have greater negative heave than the boat without foils in this experiment.

4.5 DISCUSSION

The towing tank testing of the three main options for foil-assisted monohulls has been conducted for a range of upwind and downwind conditions. The purely hydrodynamic analysis provided experimental evidence of the effectiveness of hydrofoils and yielded a number of important results.

Firstly, the induced drag factor appears a more sensible method to assess the ideal configuration compared to the effective draft, as the former enables to identify the typical operating range of the yacht in terms of side force, whereas the effective draft could suggest an erroneous interpretation.

Then, to generate a given side force, the boat without foils will create a lesser resistance than any of the three geometries tested.

The Dali-Moustache foil is the only arrangement that creates significant side force without leeway. This is surprisingly not the case for the Chistera and was expected for the DSS. Moreover, below a Froude number of 0.50, the vertical lift is not sufficient for the displacement to be reduced. Past that Froude number, the DSS develops the most vertical lift (in addition to the one generated by the vessel reaching semi-displacement mode). Moreover, at any Froude number, the Dali-Moustache performs worse than the configuration without foils.

Finally, when investigating the effects of an increased AoA, the Chistera appears to respond better to a higher angle. The DSS, however, operates best with no AoA, as the vessel's trim allows the section to operate very close to its ideal lift/drag ratio. Finally, the Dali-Moustache functions optimally at a moderate AoA upwind (8°) and a higher AoA downwind (16°). A varying angle of incidence can, therefore, be beneficial on a Dali-Moustache foil to boost either the side force or the heave.

Overall, building on the experiments undertaken and hydrodynamic data gathered, it appears that, for foil-assisted monohulls, no resistance advantage over a design without foils could be achieved, thus demonstrating their inefficiency under the present test conditions and inherent limitations, namely the pure hydrodynamic efficiency of foil-assisted monohulls.

Nevertheless, the increasing presence of hydrofoils in offshore racing yachts and now cruising superyachts suggest there are indeed strong advantages. These observations and present experimental results, therefore, call for further work to tackle the stability and performance aspects, and identify where the benefits of foils truly are, so that their design can be better refined, and the most suitable configuration selected for a vessel's operating profile.

5. STABILITY

5.1 INTRODUCTION

The tests undertaken for the purpose of quantifying the added dynamic stability were performed in conditions representative of upwind ($\theta = 20°, \lambda = 3°$) and downwind ($\theta = 10°, \lambda = 0°$) sailing. The slightly reduced leeway in the upwind condition was dictated by the free-to-heel setup that could not cope with the larger side force generated for higher leeway angles.

These experiments featured a new aft position for the Dali-Moustache and Chistera. This would not allow the foil to fit within the maximum hull width as intended in the previous experiment, but could provide greater stability and thus be of interest for racing crafts.

It must also be emphasized that the heel angles quoted correspond to the dynamic angle adopted by the yacht without hydrofoils, tested at a given speed. This, therefore, required trial and error to assess, for each Froude number, the transverse ballast location and inherent starting static heel angle, so that the vessel would reach the desired dynamic angle once towed.

5.2 RIGHTING MOMENT

For this particular test campaign, the vessel was not constrained in its heel angle. The righting moment provided by each foil was quantified from the change in heel angle measured. Firstly, an inclining experiment was conducted on the model fitted onto the towing tank carriage to establish the position of the centre of gravity. This information was then combined with the model geometry in a large angle stability analysis to determine the righting moment at every heel angle. The difference between the righting moment with and without foils, therefore, gives the dynamic contribution to the stability of the yacht. The results, in the form of the added percentage of righting moment compared to the hull fitted with a keel and bulb only, are presented in Figures 7 (a) and 7 (b) for upwind and downwind respectively.

The Dali-Moustache foil in the aft position proved to be the best in terms of generating righting moment at any heel angle; it must be noted that in certain cases, the foil was able to bring the boat back beyond the upright and into negative heel; those results should, therefore, be considered with care. Upwind, the performance of the Dali-Moustache is matched by the DSS, the latter suffering from ventilation issues due to the proximity with the free surface at the highest Froude number, hence the sudden decrease in righting moment. On the other hand, the Chistera foil only provides minor improvements downwind and reduces the dynamic stability in its forward position upwind.

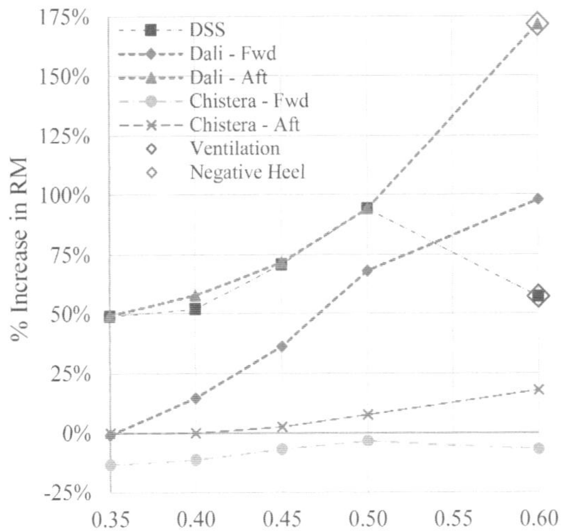

a)

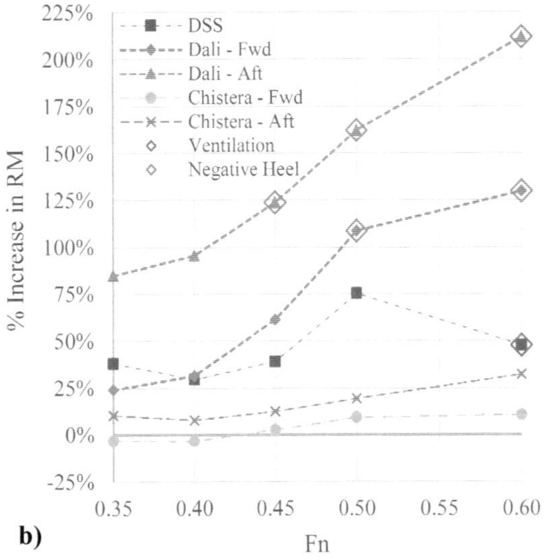

b)

Figure 7: Added righting moment provided by the foils (a) upwind and (b) downwind.

For both the Dali-Moustache and Chistera foils, the aft position is far better in term of the contribution to added dynamic stability, primarily because it is located further away from the centerline. The aft location should, therefore, be preferred, provided practical considerations do not dictate a forward position, for example, so that the retracted foil fits within the overall breadth for mooring purposes, a vital aspect for cruising vessels.

Finally, the experiments demonstrated that a yacht or superyachts subject to a given heeling moment onto which a suitable foil is added will benefit from a drastic increment in stability, with however no decrease in drag.

5.3 RESISTANCE

The results proved very consistent with the original hydrodynamic efficiency experiment in that the lowest resistance is always achieved without foils. Indeed, despite the vertical heave (only significant from $Fn = 0.5$), the reduced displacement and wetted surface area are never sufficient to overcome the added resistance and induced drag of the foil. Notably, the configurations providing the most righting moment, namely the Dali-Moustache in both conditions and the DSS upwind, also have the most drag, as shown upwind in Figure 8 (a) and downwind in Figure 8 (b).

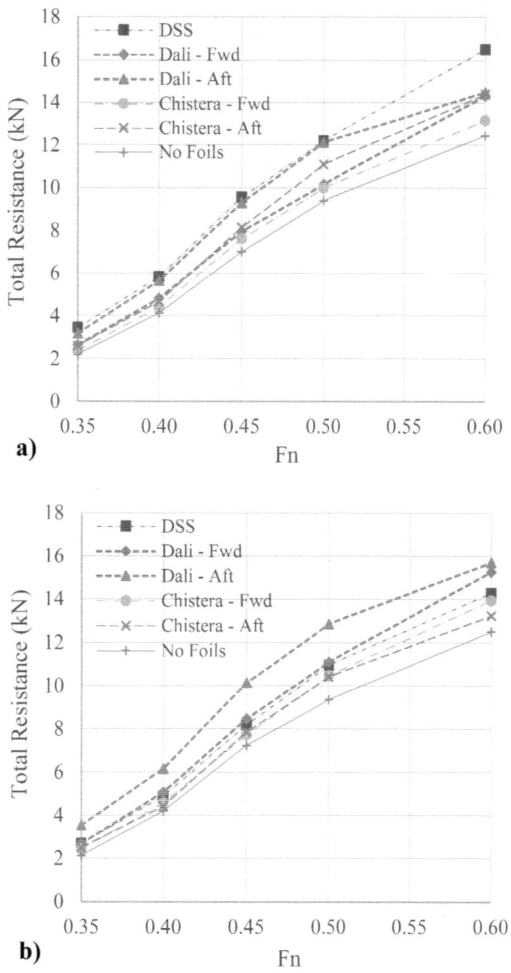

Figure 8: Total resistance (a) upwind and (b) downwind.

The resistance is a vital factor in the performance of a yacht: for a given drive force, the vessel with the least resistance will be the fastest. Furthermore, for upwind sailing, side force is critical. Interestingly, the Dali-Moustache generated the most side force in its aft position, while the Chistera did so in its forward position. Remembering that IMOCAs feature a Dali-Moustache aft and the Figaro Bénéteau 3 has a Chistera forward, the fact that each one is located in the position developing maximum side force could imply this is the parameter designers have been trying to improve for optimized performance. For the Dali-Moustache, the aft location is also the best position to generate stability. It is however not the case for the Chistera foil, which could suggest its primary objective is not to improve the stability but provide additional side force. Ultimately, this would allow the vessel to sail closer to the wind, and thus travel a shorter route into the wind. This will be further analysed in Section 6.4, taking into account the effect of the leeway angle on upwind performance to compare the theoretical sailing times on the water around an upwind race course.

5.4 DISCUSSION

The investigation into the ability of hydrofoils to create dynamic stability provides an insight into the added stability due to various configurations and positions. The present work is consistent with the earlier findings relative to the hydrodynamic efficiency and provides tangible arguments regarding the influence of foils on stability.

5.4 (a) Dynamic Stability System

The DSS demonstrated a very effective contribution to the righting moment upwind. In this particular instance, for $\theta = 20°$ and $\lambda = 3°$ at $Fn = 0.35$, *i.e.* a typical upwind sailing condition at the upper end of upwind speeds, a reduction in heel angle of 4.73° was measured. This is in agreement with the findings of Welbourn, inventor of the DSS: *"I figured out a simple rule way back at the very beginning of all this, and you should be looking to optimize the foil for about 5 degrees heel equivalent of RM at the top end of typical upwind speeds"* (Welbourn, personal communication, 14 December 2017).

5.4 (b) Dali-Moustache

The Dali-Moustache in the aft position (where it is found on racing yachts), revealed the best ability to add stability and reduce heel angle. While the DSS is limited by ventilation issues due to the proximity with the free surface, the Dali-Moustache proved to be able to bring the vessel past upright, thus demonstrating its efficiency at low heel angle, characteristic of downwind sailing. Since this configuration is seen on IMOCAs, primarily optimized for downwind, it is no surprise to see it perfectly suited for this point of sail. Moreover, the added stability explains the reason behind the latest generation of IMOCAs being narrower (Beyou, 2017): with the tremendous dynamic stability provided by the foils, the form stability due to the width of the vessels can be decreased, in turn resulting in a yacht with lower wetted surface area, but also a lighter weight thanks to the diminished size.

5.4 (c) Chistera

The Chistera exhibited a greater contribution to stability in its aft rather than forward position, the actual amount however being the lowest compared to other configurations. The advantages of the Chistera in the forward position, where it is found on the Figaro Bénéteau 3, are a lower drag and greater side force upwind. This would, therefore, suggest its design is targeted at a faster boat, able to sail with less leeway upwind. This would also explain the previously not understood reason for the new Figaro Bénéteau 3 featuring a heavier and deeper keel despite the foils (Dewavrin, 2018). This can now be explained as compensating with weight stability for the minimal increase in dynamic stability. Practical considerations also drive the forward location of this foil, as discussed in Section 7.3.

5.4 (d) Findings

The DSS appeared to be most suited to upwind sailing, and a very similar reduction in heel angle compared to the rule of thumb developed by the DSS' inventor has been observed. Downwind, or at low heel angles, the proximity to the free surface negatively affects this configuration, with limited stability gains.

The Dali-Moustache is creating the most righting moment in all conditions, especially in its aft position. This justifies its presence further aft on racing yachts, as well as why the latest generation of IMOCAs can afford to reduce the form stability of the hull, now mostly relying on the tremendous dynamic stability of the foils.

The Chistera foil in its forward position proved less efficient stability-wise. This would, however, explain why the new generation with hydrofoils features a deeper and heavier keel. Nevertheless, with a greater side force and lower drag upwind, it could be suggested that this is where the benefits of this configuration reside.

With the knowledge of hydrodynamic efficiency of these foils and the characterization of the added righting moment, the understanding of hydrofoil-assisted monohulls has been strongly extended. The final element to be ascertained is the overall impact on performance. Indeed, added stability will increase the power to carry sail, but it has also been shown to enlarge the total resistance. Similarly, greater side force will make for a shorter distance upwind, but again at a cost in terms of induced drag. Consequently, the development of a velocity prediction program able to capture the various behaviours of the foils will be tackled, to eventually establish their significance to the overall sailing speeds.

6. PERFORMANCE

6.1 INTRODUCTION

While the hydrodynamic and stability influence of the various foils have been quantified, the sailing speeds remain to be assessed. Velocity prediction programs have been successfully employed for the comparative performance of racing yachts (Thomas & Souppez, 2018) as well as cruising vessels (Guell & Souppez, 2018), but current commercial packages do no account for the effect of hydrofoils. Consequently, a dedicated VPP was developed (Borba Labi, 2019).

6.2 METHODOLOGY

The three degrees of freedom VPP (surge, sway, roll) relies on hydrostatics and stability input. On the other hand, the Offshore Racing Congress (ORC) methodology (ORC, 2017) was adopted to quantify the sail forces, and the DSYHS regression equations provided the hydrodynamic hull resistance model (Keuning & Katgert, 2008). For the hydrofoils, Glauert's biplane theory corrected for proximity with the free surface was utilised (Daskovsky, 2000), implementing correction coefficients based on the towing tank results obtained previously. Indeed, the empirical nature of the mathematical model representing the forces generated by the foils does not account for all the variables, hence the addition of an efficiency factor to bring the theoretical prediction in line with the experimental results for the various degrees of freedom considered.

The heave has been neglected in this instance has it was previously shown to be beneficial for typical sailing Froude numbers (see Section 4.4), and a significant reduction in displacement would not be expected on superyachts, as it is on some of the small and light racing crafts.

The VPP developed, having the architecture depicted in Figure 9, was first validated for non-foiling vessels against commercial packages to demonstrate its validity and suitability, before analysing the behaviour of hydrofoiling yachts.

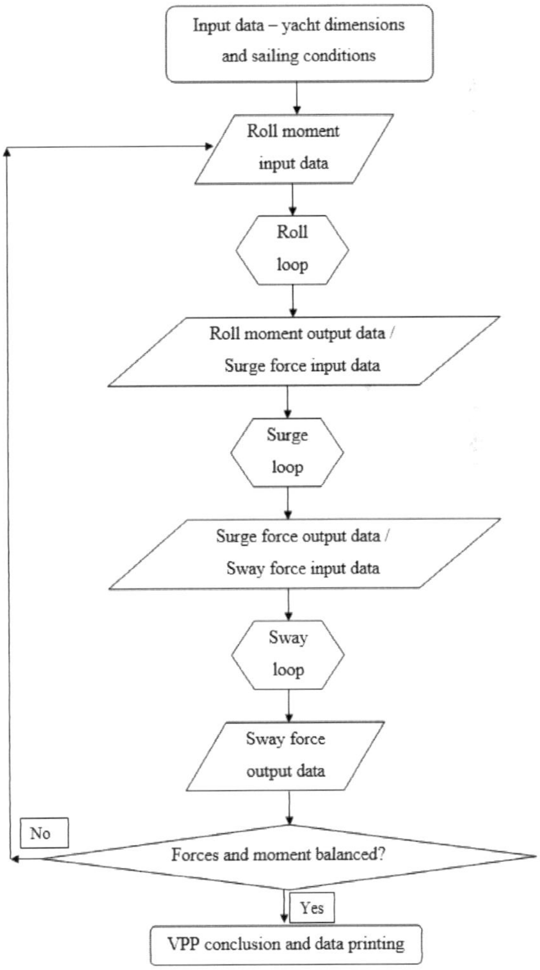

Figure 9: Structure of the VPP (Borba Labi, 2019).

Due to its nature and underpinning theory, this VPP is solely intended for foil-assisted yachts (*i.e.* not fully flying), and is best utilized at an early development stage, for the purpose of performance assessment, as illustrated in Section 6.3, but also hydrofoil design optimisation later highlighted in Section 6.4.

6.3 RESULTS PRE-OPTIMISATION

The initial assessment was conducted on the foil geometries as tested in the towing tank. The intention being to translate the experimental measurements into a quantifiable performance on the water.

6.3 (a) Upwind

In upwind conditions (mainsail and jib) for a low wind speed of 8 knots, the overall performance of each foil is very similar, with an increasing advantage in boat speed. The main difference, however, lies in the heel angle adopted by the vessel (Figure 10), highlighting the added stability of the Dali-Moustache, resulting in a lower heel angle.

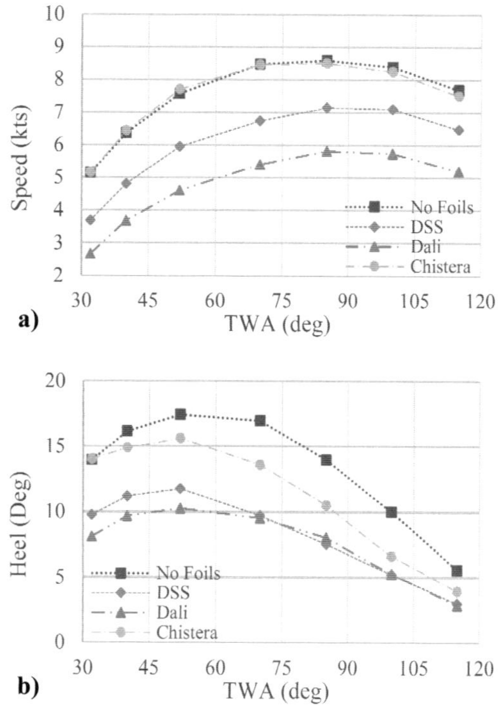

Figure 10: Upwind boat speed (a) and heel angle (b) in 8 knots true wind speed.

In low wind speeds, the small angles of heel are not yet optimum for the DSS, which provides greater stability further away from the free surface as the boat heels over more in stronger wind. This is reflected in Figure 11, where the vessel fitted with a DSS has the lowest heel angle. In terms of performance, greater differences are now shown, with the Chistera achieving higher velocities.

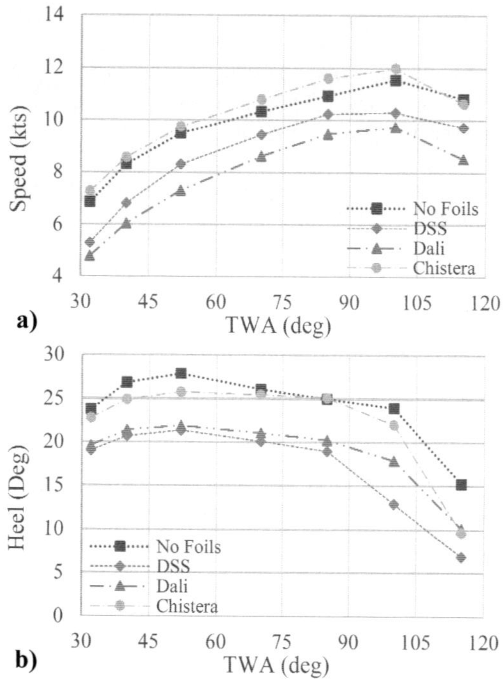

Figure 11: Upwind boat speed (a) and heel angle (b) in 16 knots true wind speed.

In those conditions, the pure boat speed is only superior to the vessel without foils when fitted with a Chistera. The DSS and Dali-Moustache do however contribute to a much lower heel angle, which could be seen as more suitable in cruising conditions for comfort.

6.3 (b) Downwind

In the downwind case (mainsail and spinnaker), the results are less sensitive to the wind speed. Both low (Figure 12) and high (Figure 13) wind speeds depict identical trends, with the Chistera performing best, but at a higher heel angle. Here again, only the Chistera proved able to surpass the boat speed of the yacht not fitted with hydrofoils.

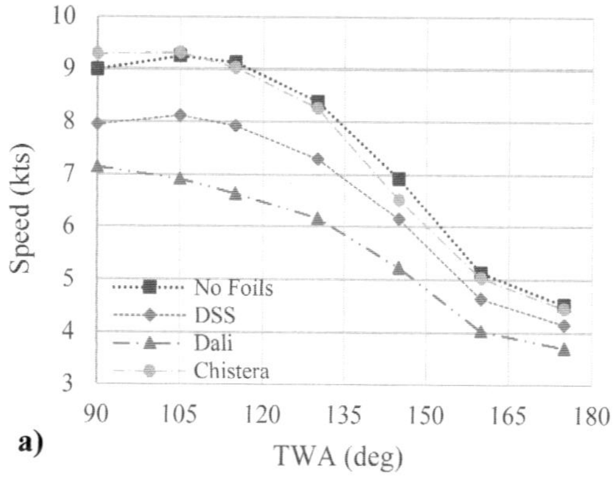

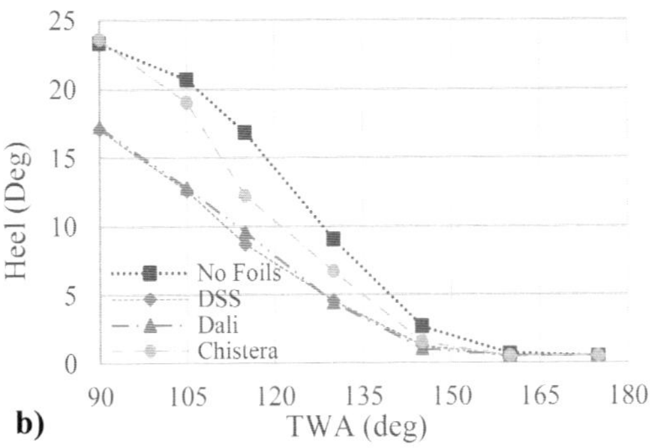

b)

Figure 12: Downwind boat speed (a) and heel angle (b) in 8 knots true wind speed.

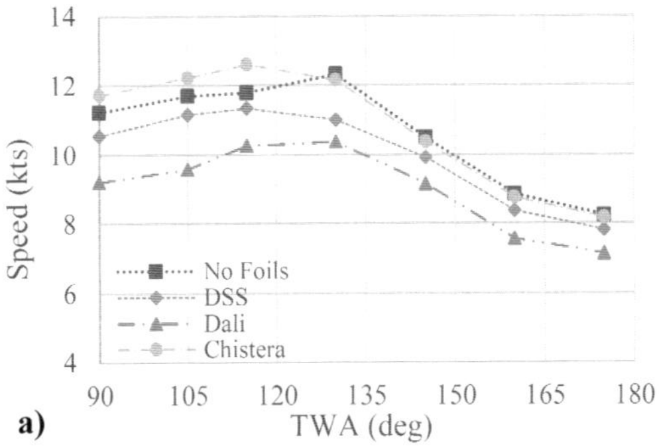

a)

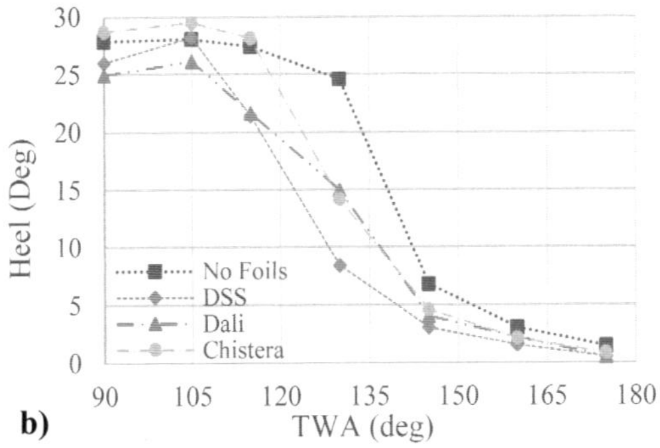

b)

Figure 13: Downwind boat speed (a) and heel angle (b) in 16 knots true wind speed.

Those results are however not sufficient to ascertain the Chistera foil as the best performing one in all conditions. On the one hand, an identical vessel has been considered, when added sail area could, for instance, be fitted on a boat equipped with Dali-Moustache or a DSS due to the significant added righting moment. For a given heel angle, those two configurations would have more power and thus achieve better speeds. In addition, the leeway angle would be considerably reduced if the yacht was fitted with a Dali-Moustache or Chistera. On the other hand, the comparison presented is for a given foil design that has not been optimised. The original towing tank tested geometries resulted from a parametric analysis to ensure their representative nature, but in light of the new experimental findings and the VPP created, their design can be refined. Consequently, the specifications of each hydrofoil will be altered to achieve an optimum geometry before re-assessing the performance.

6.4 RESULTS POST-OPTIMISATION

The VPP created permits to conduct a parameter study of the hydrofoil geometries with the aim of maximising performance. The design optimisation was targeted around some key features, namely: the span, aspect ratio, the angle of the foil to the hull and angle between its two part for the Dali-Moustache and Chistera.

The results showed that very little improvement could be made on the initial Chistera geometry that already appears to be extremely efficient; this explains its superiority in the previous presented section. On the contrary, significant performance optimisation could be achieved with both the DSS and Dali-Moustache. The gains between the original tank tested versions and the VPP optimised ones are presented for 16 knots of wind upwind and 18 knots downwind in Figure 14.

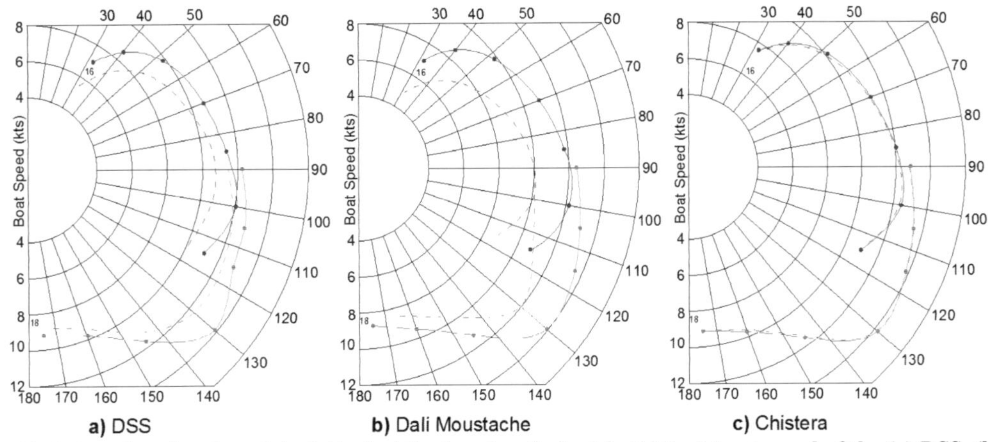

Figure 14: Polar plots for the original (dashed line) and optimised (solid line) boat speed of the (a) DSS, (b) Dali-Moustache and (c) Chistera.

Lastly, as all VPPs, the presented one should be considered qualitatively, allowing to compare the performance of various boats, rather than quantitatively. Indeed, although similar results between VPPs and sea trials can be achieved (Souppez, 2014), there is now evidence to suggest the force coefficients employed as part of VPP models and originating from wind tunnel tests could be flawed (Souppez *et al.*, 2019)

Nevertheless, this illustrates the crucial importance of the qualitative VPP at early design stages. Having reached an optimal geometry for each configuration, the performance of the three vessels could be compared again, yielding very interesting results. Indeed, with the optimised hydrofoil designs, virtually no differences in velocity or heel angle were present, as illustrated for an upwind case in Figure 15.

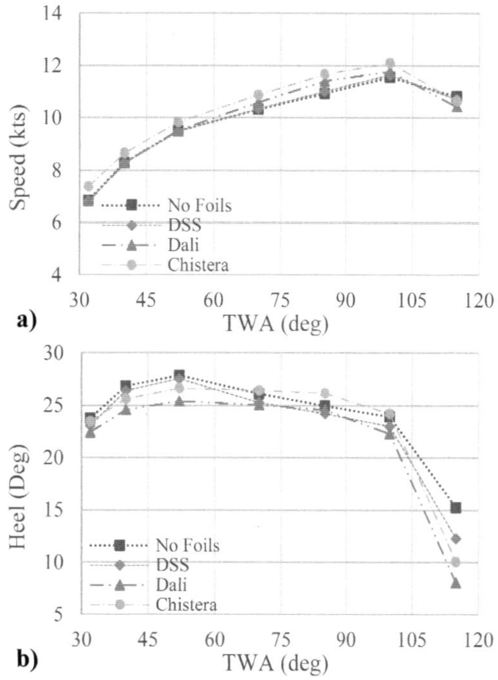

Figure 15: Upwind boat speed (a) and heel angle (b) in 16 knots true wind speed for optimised hydrofoils.

Therefore, it appears that provided the design of the hydrofoil is optimised, similar velocities can be achieved, irrelevant of the actual configuration employed on the vessel. However, the comparison must also consider the leeway angle, with a strong difference between the arrangements not creating side force (no foils and DSS) and those that do (Dali-Moustache and Chistera), the later having a much smaller leeway angle, as quantified in Figure 16.

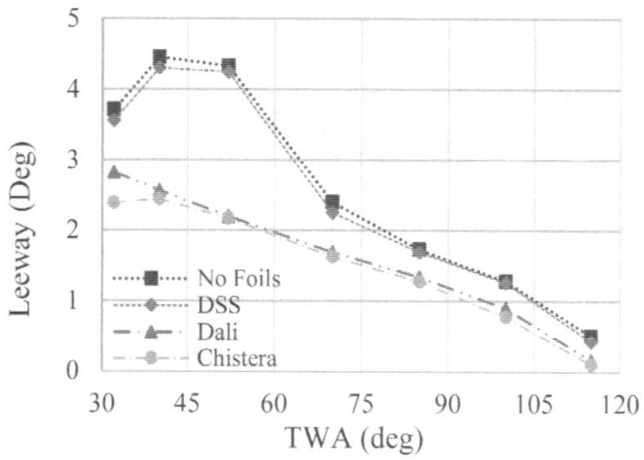

Figure 16: Upwind leeway in 16 knots true wind speed for optimised hydrofoils.

With this information, for a one nautical mile upwind course, the Chistera would be the first at the mark, followed by the Dali-Moustache 27.2 seconds behind, and then the vessel without foils and the DSS, respectively 44.3 and 44.7 seconds later. This therefore provides a clear comparison of the actual performance on the water of the various designs.

6.5 DISCUSSION

An empirical VPP tool able to account for the effect of hydrofoils was devised to quantify the sailing performance and demonstrate the significant impact of their optimisation on boat speed for foil-assisted monohulls. One of the key findings is that, for designs that have been ameliorated, there is no configuration superior to another in terms of velocity, including a yacht without foils. However, there are strong differences in terms of performance on the water, with a clear ranking between the different options.

It is worth noting that this study assumed that the design of the yacht remains constant. In practice, the amount of sail area or hull shape could be refined based on the specificities of each hydrofoil, as tackled in Section 7.5.

Nevertheless, the choice of which option to be installed on a yacht or superyacht is also subject to practical considerations.

7. PRACTICAL DESIGN CONSIDERATIONS

7.1 INTRODUCTION

Since the same performance can be attained for any of the three hydrofoil configurations investigated, the practical design considerations are vital factors to consider. These primarily revolve around minimising the loss of internal volume, ease of mooring and preventing marine growth on the foils. These also supplement elements normally considered as part of the development process, such as issues associated with cavitation and ventilation, or the structural loads, although these are currently beyond the scope of structural design regulations (Souppez, 2018).

7.2 INTERNAL VOLUME

Accommodation volume is always limited and must be maximised, even on the largest mega yachts. As a result, the intrusiveness of the hydrofoils should be minimized. To that effect, the DSS is the easiest to fit, as it can easily be concealed under the floorboard, even on small crafts (Guell & Souppez, 2018), thus having very little impact on the interior volume. The Chistera, and to a greater extent the Dali-Moustache, however, induce in a much larger loss of volume.

Of course, the physical size of the yacht itself play a large role. At present, only the DSS has been featured on superyachts, which would appear a sensible solution to avoid the loss of internal spaces, but also for mooring and maintenance reasons.

7.3 MOORING

An additional factor to consider in implementing hydrofoils onto superyachts is the ability to fully retract it within the overall breadth of the boat for mooring. Not only is it more expensive and harder to find suitable berth for a wider vessel, but hydrofoils are fragile, and should be protected.

A hydrofoil such as the Dali-Moustache protrudes beyond the overall beam of the vessel, thus requiring a larger berthing space as well as suitable protection for the hydrofoils. This also implies the deck edge will be further away from the quayside, which could represent a loading/unloading issue.

Consequently, it could be seen beneficial to prevent this situation. In the case of the DSS, the foil retracts and can be stored within the breath of the hull. This generally governs its position further aft, where a greater breadth is available. On the other hand, the Chistera, which still practically protrudes once retracted, is located forward, where the boat is narrower so that the outer extent of the foil remains within the overall breadth.

Irrelevant of the size of the vessel, preventing the hydrofoils from sticking out of the hull's overall beam in the harbour is to be considered by the designer. This can also help prevent marine growth and thus minimise maintenance by keeping the foil dry.

7.4 PREVENTING MARINE GROWTH

The performance gains obtained from the hydrofoils rely on a high lift to drag ratio. Unfortunately, the development of marine growth on its surface sharply hinders its effectiveness. The hydrofoil's surface should, therefore, remain smooth; the easiest way being to keep the hydrofoil dry when not in use.

On racing yachts such as the IMOCAs the foils will remain submerged even when retracted. This issue is alleviated by the fact that racing yachts will be regularly cleaned, and a high level of maintenance will be available ahead of the race start. A racing class such as the Figaro Bénéteau 3, equipped with Chistera foils, would also be expected to benefit from this.

For a more leisurely application of hydrofoils, which represents most of the sailing yacht industry, a more maintenance-free solution should be reached. Here again, the ability of the DSS to fit within the hull shell and the Chistera being mostly outside of the water when retracted provide strong practical arguments for their use.

7.5 DISCUSSION

The selection and design of a given hydrofoil configuration should consider all aspects, from the performance to the more practical elements. With the ability to fit within the hull's overall breadth, be kept away from the environment when retracted and minimal loss of internal volumes, the DSS appears as an easy system to install and has currently been the most widespread form of hydrofoil on sailing yachts and superyachts.

Nevertheless, other hydrofoil configurations can lead to alternative design philosophies. The Dali-Moustache for instance, shown to provide the most added dynamic righting moment, has led to new hull design for the IMOCA class. Indeed, the latest vessels feature narrower hulls, with less form stability, no longer required thanks to the foils. This also diminishes the build cost and weight as the surface area and size of the craft is effectively lowered.

8. CONCLUSIONS

Extensive experimental hydrodynamic testing has been performed on three contemporary hydrofoils in order to further the knowledge of hydrofoil-assisted monohulls, for application ranging from small racing yachts to cruising mega yachts.

Firstly, the hydrodynamic efficiency investigation revealed that, despite their contribution to the vertical lift and side force, none of the tested configurations could achieve a lower drag that the hull without hydrofoils.

This prompted further work to quantify the added righting moment provided, in a free-to-heel setup. The results showed that, while significant dynamic righting moment could be created, this came at a strong drag penalty.

In order to ascertain how the previous findings, influence the overall speed of the yacht, which is of paramount importance, a dedicated velocity prediction program was developed. This tool allowed to define the comparative performance of the various foil types, but also to conduct a parameter study and refine their design. Upon optimisation of each hydrofoil, it appeared that none could provide a greater speed than the others.

Furthermore, it is interesting to note that, for foil assisted-monohulls where the foil does not a provide significant reduction in heave (particularly on superyachts where the lift force is very small compared to the vessel's displacement, and would not occur until a higher Froude number, as demonstrated experimentally), the actual boat speed remains virtually unchanged. There are however some strong benefits in terms of reducing the heel angle for comfort and leeway for performance that can be very attractive. Overall, looking at the race time on an upwind course, the Chistera would win, followed by the Dali-Moustache, and eventually the DSS, the later achieving a similar time as a yacht without foils.

Finally, the practical considerations that could influence the selection and design of the most appropriate arrangement have been outlined, revolving around the ability to retain internal volume, ease of mooring and the prevention of marine growth.

These novel findings provide new insights into the design of hydrofoil-assisted monohulls. Future work will, however, consider the impact on the design of the vessel itself. Indeed, all configurations were tested on an identical hullshape and sailplan. In practice, a narrower hull with less drag and less mass could be designed thanks to the added dynamic stability. Moreover, a greater sail area could be implemented as the power to carry sail has been increased, eventually resulting in a faster yacht.

In addition, research into the seakeeping characteristic of the vessels should be undertaken, with a potential reduction in motions experienced for greater comfort and lower structural loads.

9. ACKNOWLEDGEMENTS

The authors would like to greatly acknowledge the support of the following organisation and individuals:

- The West Pomeranian University of Technology, Poland, the University of Rostock, Germany and Solent University, UK, cooperating in the framework of the EMship+ Erasmus Mundus Joint Master's Degree programme, coordinated by the University of Liège, Belgium.

- The Bénéteau Group and the ENSCBP University, France, for the funding and supporting the stability investigation.

- J. Cunningham-Burley for his support during the manufacturing of the model, C. Kitching for his assistance with the experimental testing, and G. Kay and H. Welbourn for their expert insight about DSS hydrofoils.

10. REFERENCES

1. Abbott, I. H. & Doenhoff, A. E., 'Theory of wing sections, Including a Summary of Airfoil Data', *New York: Dover Publications Inc*, 1959.

2. Beaver, B. & Szeleczky, J., 'Full Scale Measurements on a Hydrofoil International Moth', *The 19th Chesapeake Sailing Yacht Symposium*, Annapolis, USA, 2009.

3. Beyou, J., 'On Peut Envoyer Fort Dans La Grosse Brise!', *Voiles et Voiliers,* 2017.

4. Binns, J. R., Ashworth-Briggs, A., Fleming, A., Duffy, J., Haase, M., & Kermarec, M., 'Unlocking Hydrofoils Hydrodynamics with Experimental Results', *The 4th International Conference on Innovation in High Performance Sailing Yachts*, Lorient, France, 2017.

5. Borba Labi, G., 'Velocity Prediction Program Development for Hydrofoil-Assisted Sailing Monohulls', EMship Master Thesis, *Solent University*, 2019.

6. Campbell, I., Owen, M., & Provinciali, G., 'Dagger-board Evaluation for an IMOCA 60 yacht', *Ocean Engineering*, vol. 90, 2014.

7. Daskovsky, M., 'The hydrofoil in surface proximity, theory and experiment', *Ocean Engineering*, 27(10), pp. 1129–1159, 2000.

8. Dewavrin, J. M. M.-A., 'Design Of A Cruising Sailing Yacht With An Experimental Fluid Dynamics Investigation Into Hydrofoils', EMship Master Thesis, *Solent University*, 2018.

9. Dewavrin, J. M. M.-A and Souppez, J.-B. R. G., 'Experimental Investigation Into Modern Hydrofoils-Assisted Monohulls: How Hydrodynamically Efficient Are They?' *International Journal of Small Craft Technology*, Part B2, Vol 160, pp111-120, 2018.

10. Guell, A. and Souppez, J.-B. R. G., 'Combining Modern Hydrofoils with Wooden Classic', *British Conference of Undergraduate Research*, Sheffield, UK, 2018.

11. ITTC, 'ITTC – Recommended Procedures – Fresh Water and Seawater Properties', *International Towing Tank Conference*, 2011.

12. ITTC, 'ITTC Quality System Manual - Recommended Procedures and Guidelines - Guidelines - General Guideline for Uncertainty Analysis in Resistance Tests', *International Towing Tankk Conference*, 2014.

13. ITTC, 'ITTC Quality System Manual - Recommended Procedures and Guidelines - Procedure - Ship Models', *International Towing Tank Conference*, 2017.

14. Jackson, P. & Hawkins, P., 'The Importance of Froude and Reynolds Numbers in Modelling Downwind Sails', *13th Australasian Fluid Mechanics Conference*, Melbourne, Australia, 1998.

15. Keuning, J. A., & Katgert, M., 'A Bare Hull Resistance Prediction Method Derived from the Results of the Delft Systematic Yacht Hull Series Extended to Higher Speeds', *Delft University of Technology*, 2008.

16. Kitching, C. B. F., 'A Luxury 24m Foil Assisted Round The World Cruiser Racer', *Solent University*, 2018.

17. ORC, 'ORC VPP Documentation', *Offshore Racing Congress*, 2017.

18. Sacher, M., Durand, M., Berrini, E., Hauville, F., Duvigneau, R., Le Maitre, O., & Astolfi, J. A., 'Flexible Hydrofoil Optimization for the 35th America's Cup Constrained Ego Method', *Innov'Sail - The Fourth International Conference on Innovation in High Performance Sailing Yachts*, Lorient, France, 2017.

19. Souppez, J.-B. R. G., 'Development and Validation of a Computational Fluid Dynamics Hydrodynamic Model of the Stewart 34 for Velocity Prediction Program Applications', Master Thesis, *The University of Auckland*, 2014.

20. Souppez, J.-B. R. G., 'High Performance Racing Yachts: An Experimental Comparison of the Latest Hydrofoil Configurations', *UK Fluids Conference*, Manchester, UK, 2018.

21. Souppez, J.-B. R. G., 'Structural Design of High Performance Composite Sailing Yachts under the New BS EN ISO 12215-5', *SNAME Journal of Sailing Technology*, issue 2, pp 1-18, 2018.

22. Souppez, J.-B. R. G., Arredondo-Galeana, A. & Viola I. M., 'Recent Advances in Downwind Sail Aerodynamics', *The 23rd Chesapeake Sailing Yacht Symposium*, Annapolis, Maryland, United States, 2019.

23. Thomas, J. and Souppez, J.-B. R. G., 'Comparative Performance Prediction of Historical Thames A Rater Class Designs', *Historic Ships*, London, UK, 2018.

11. AUTHORS BIOGRAPHY

Jean-Baptiste R. G. Souppez is the Senior Lecturer in Yacht Design and Composite Engineering at Solent University, teaching on the prestigious *BEng (Hons) Yacht and Powercraft Design, BEng (Hons) Yacht Design and Production* and *MSc Superyacht Design*. He contributes to the European Master in *Integrated Advanced Ship Design* (EMship+) as a Visiting Professor and Research Supervisor, and is also the UK Principal Expert in Small Craft Structures, in charge of representing the interests of the British Marine Industry in the development of international structural regulations (BS EN ISO 12215). His research in fluid dynamics features twisted flow wind tunnel, towing tank, wave and current flume, particle image velocimetry, laser doppler anemometry, and full-size instrumented testing, as well as a range of numerical methods.

Juliette M. M.-A. Devawrin is currently a Graduate Design Engineer at Gurit Composite Engineering, UK. She is working on pre-studies and structural engineering solutions for composite structures in glass fibre and carbon fibre reinforced polymers on various projects such as sailing and motor boats, superyachts, vehicles and transportation, architecture and renewables. She is also conducting an experimental research on naval composite structures for Gurit internal development in the scheme of her graduate programme. During her studies on the EMship+ master in Ship Design, she conducted experimental research in fluid dynamics at Solent University, featuring modern hydrofoils-assisted monohulls hydrodynamic efficiency.

Florian Gohier presently occupies the position of apprentice engineer in composite engineering between the ENSCBP (Bordeaux Graduate School of Chemistry, Biology and Physics) and BJ Technology, Bénéteau Group, France. He is responsible for the structural calculations of hulls, hull liners, decks, deck liners and composite components manufactured by the Bénéteau Group.

Gaetan Borba Labi is currently a mooring engineer at Principal Power, Portugal. His responsibilities include the analysis and design of mooring systems for offshore platforms used in the Eolic industry of energy generation. His previous experience features numerical modelling of fluid dynamic problems and development of tools for the velocity prediction of hydrofoils-assisted monohulls.

DEVELOPING THE CONSTELLATION UNMANNED SURFACE VEHICLE PROTOTYPE

R G Shaw, New Zealand Maritime School, Manukau Institute of Technology; Shaw Yacht Design, New Zealand

SUMMARY

Vessel navigation through the use of reliable nautical charts is vital to ensure safe passage. Having the ability to complement this with the ability to see accurately both what lies beneath the water and ahead of a vessel in real time provides an added dimension of safety and certainty. Combining the high value and deep draft of modern super and mega yachts with a penchant for exploration creates a tension which can test the availability and accuracy of navigation charts.

The development of a 2 metre remote-operated and autonomous-capable surface vehicle with ultra-high-resolution echo-sounder capability provides a unique solution which can open up considerable exploration capability. The ability to produce ultra-high-resolution images and relay these back to the mother vessel in real time enables precise seabed mapping, providing safer navigation. Additionally, it has the capability to deploy personal water-safety devices in situations such as a man-overboard incident.

1. INTRODUCTION

There is a constant drive for improvement through innovation across all forms of transport technology. This quest for continued improvement is strong in the field of marine Unmanned Surface Vehicle (USV) design. The design and development of this type of craft is driven by a constant need for refinement and innovation to meet changing demands.

This paper explores the development of this remote vessel through the stages of concept development, design, manufacture and performance validation. The vessel's design development evolution is explored and assessed against operational and performance objectives. Prototype vessel performance is validated and presented to verify performance and accuracy of data collection. The resulting vessel provides unique capabilities which can enhance the safety of larger vessels by complementing contemporary navigation systems.

The focus of this paper is on the development of a 2 metre autonomous surface vessel. The vessel can be operated either remotely or autonomously and has the capability to provide high-quality seabed mapping. With advancements in technology both providing the ability to build lighter, stronger structures and enabling lighter and more compact power plants, a great opportunity exists to reconsider how we, as designers, can create a more efficient USV.

Typically, advancements in light-weight composite construction have been applied to craft seeking to achieve higher performance and higher speeds. In this application the advancements in materials and process have been adapted to provide a light, strong and durable vessel which is both versatile and optimised to operate at slow speeds.

This paper considers the objectives for a remote vessel with high-definition survey capabilities in the context of how this technology can be deployed to provide benefits to super and mega yacht operators. The outcome is a high-quality, high-technology and higher-specification USV that is light, versatile and highly efficient. Through a combination of refined hull form, modular design and advanced construction technology, the combination of higher performance with lower-speed craft is explored.

2. DESIGN CRITERIA AND OBJECTIVES

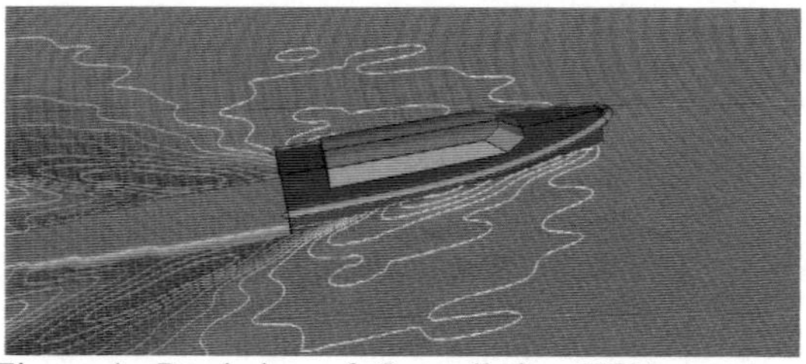

Figure 1: Rendering of Constellation USV prototype being deployed.

For this new design it was important to frame its priority objectives. This is not intended to be a radical design that signals a major departure from current form or function. The purpose is to develop a design which makes best use of current advancements in design, materials and manufacture methods. Through the use of the best materials and techniques available it is possible to produce an USV which enables the use of these vessels in a different context to conventional solutions.

The key objective for this design is for the vessel to have the potential to efficiently operate for extended periods at speeds considered higher than those associated with conventional displacement mode operation. In the 4-10 knot speed range the craft operates with a Froude Number of 0.4-1.2. The Constellation USV is primarily designed to carry out survey services on restricted waterways and harbours where access for conventional survey craft is difficult.

Survey services may include nautical charting; baseline, habitat and environmental mapping; and determination of hazards. These capabilities and the versatile design lend the USV to be deployed as a scouting survey drone to escort super yachts and similar craft into harbours and anchorages where accurate charting of the seabed contours is not available or reliable.

The system has a modular design; interchangeable equipment capsules are able to be removed ensuring simple and safe transport by land, sea or air. The large, 700 mm x 400 mm 'moon pool' unit allows for survey-equipment capsule payloads of up to 35 kg. Self-deploying sensors mean that operations can be undertaken with no direct human contact.

3. REVIEW OF CURRENT SOLUTIONS

Previous USV designs have tended to be characterised by low construction specifications. The base vessel designs appear to have received little attention, with the focus apparently being on the system design and survey capabilities. Weight saving is also not a primary concern; for example, the Hydra USV, designed as a remote-operated security-control vessel, is 4.2 m long and weighs 400 kg. [1]

While these existing USVs achieve their objective of providing a platform for the survey equipment they carry, they leave considerable scope for design improvements. With these improvements realised, the potential for a broader scope of activities is possible.

The use of foils on small performance craft is a question that is often raised. The application of lifting foils is increasingly becoming relevant, with this technology being applied to various designs to good advantage. However, for a vessel operating in the speed range being considered in this paper, it is recognised that lifting foils would not offer a significant performance enhancement. Certainly, when considered together with the added expense and complexity of such systems, it is not an attractive option for a vessel of this type.

4. DESIGN DEVELOPMENT

A defining feature of this design is the moon pool removable module. This provides a great deal of versatility to the vessel, enabling for swift repurposing. By containing all of the sounding equipment on this module separate from the control, power and thrusters, the sounding equipment can be removed or replaced in a few minutes and the vessel be redeployed.

Figure 2: Underside view, showing R2 sonic 2020 sensor in position in the moon pool.

The hull form is optimised for minimum resistance at normal operating speeds while still maintaining adequate stability and enough internal volume to accommodate all of the desired systems. The hull features a fine entry, flat mid-section and twin bilge keels aft for directional stability.

The light weight, compact design and potential to remove the moon pool component and its containing media provides ease of transportation for the vessel by many modes, including the ability to transport the vessel in a backpack arrangement.

In an effort to maximise the efficiency of the design, the vessel has been developed borrowing aspects of design from the rigid inflatable boat genre. The original concept incorporated an inflatable gunwale tube, but due to the scale of the craft this evolved into a CNC-machined solid EVA foam member, finished and painted and bonded to the hull with contact adhesive. This adds to its range of stability. An additional benefit is that it plays an important role in keeping the deck of the craft dry, preventing wave slop and spray reaching the equipment aboard.

Figure 3: CAD renderings of the USV prototype.

5. CONSTRUCTION

A critical component of this USV design is the weight reduction and durability achievable through the utilisation of building in carbon composites. The use of advanced composites for large-scale construction was introduced by the aerospace industry since the 1960s, and migrated to the field of boatbuilding in 1980s. An early example was New Zealand America's Cup challenger KZ7, built in 1986.

Advanced composite construction has been utilised and developed since then, in the construction of racing yachts, both mono and multihull, and super yachts. *Hetairos III*, at 66.4 metres LOA, is the largest super yacht so far built in carbon composite. [2]

Improvements in construction techniques, including resin infusion and the use of pre-impregnated laminates has made composite construction more attractive and practical, increasing quality and reliability. The cost of carbon composite laminate has also dramatically reduced since its introduction to the boatbuilding industry.

The use of composites eliminates the traditional limitations on producing complex geometric shapes. This gives the designer unrestricted potential to explore more practical or aesthetically pleasing geometries.

From a maintenance point of view, well-engineered and manufactured composite structures are susceptible to the corrosion risks prevalent in the marine environment and are therefore longer lsating. Composite structures also have low fatigue rates and good resistance to chemical and electrolytic corrosion.

Using earlier composite construction techniques, a timber plug needed to be built and finished to a suitable standard so the structure could be laminated over it. This was labour-intensive and added to the build time and workforce requirements. However, more recent advancements in CAD-CAM technologies, larger CNC machining capabilities and the expanding capabilities of composite boatbuilders mean custom boats can be built from machined female moulds, meaning building one-off composite boats is now a viable option. [3]

For this vessel the structural surfaces are built from monolithic carbon, inside a CNC-cut female mould milled from medium-density fibreboard. Through the use of resin-infused carbon fibre laminates inside a female mould, the total dry weight of the vessel including the moon pool is 12 kg.

The hull itself was first fabricated using only the mould as the outer shape. Another mould, created for the moonpool module, was used to create a payload box and the moonpool structure from that. A section was cut from the hull and the moonpool structure glued in place. The deck of the USV was fabricated from another mould and later glued to the hull. A bulkhead seals off the forward section of the bow, increasing strength and creating an airtight space for additional buoyancy.

Figure 4: Completed USV ready for transportation and testing.

Figure 5: Carbon hull in the female mould before moulded deck was added.

6. ELECTRICAL SYSTEM DESIGN

The electrical power system on board the USV is based on a 24V DC battery system. Systems are kept separate by running sensitive equipment through a DC-DC converter. In this prototype, the operator has limited control of systems remotely, though this capability will be enhanced in the near future. An overview of the electrical system on board the Constellation USV is provided below as Figure 6.

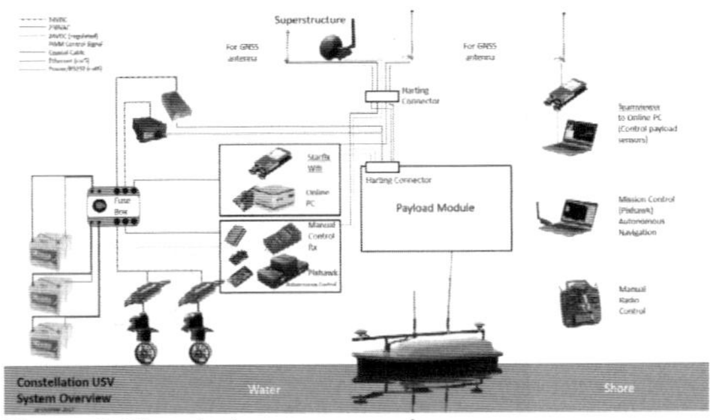

Figure 6: Electrical system overview.

A number of battery technologies are available, providing 20-30V DC during a discharge cycle. Around 1-2 kWh of energy is available from these battery packs. All systems are connected to the power source via a distribution board, where loads are fused for protection.

Four 24V, 10Ah Ni-MH batteries were the original proposed batteries for this project. These batteries did not perform as well as expected (around two-thirds of rated capacity) and the battery charging systems were inconsistent.

As a secondary pack, sealed lead acid batteries were used. These batteries are relatively inexpensive and readily available. Charging is also simple.

These are rated at 12V nominal, and 18Ah capacity – total of 216Wh each. Since they are 12V packs and the vessel requires 24V, a wiring harness setting up pairs of batteries in series was required. Batteries were marked so that they would remain as the same pair.

Initially four batteries were used, for a total of 864Wh. This yielded a longer run time than the NiMH batteries. An additional two batteries were added to the system to increase capacity to 1296Wh. This resulted in a more than 50 per cent increase in run time, as batteries are more efficient when under a lighter load. Assuming a 500W load, this six-pack system should yield a run time of 2.59 hours.

A set of K2B24V10EB Lithium Iron Phosphate batteries were also trialled. They include all electronics internally and require only a constant voltage 30V supply for charging. Over-current and reverse polarity protection are also built in to the battery.

Nominal battery voltage is 25.6V, and total capacity of 245.8Wh. A set of eight batteries totals 1966 Wh, enough for in excess of 4 hours' operation.

7. PROPULSION SYSTEM

Two Blue Robotics T200 Thrusters propel the USV. These are mounted on azimuth mounts which can be steered through about 30 degrees either way. Each thruster is controlled by a high voltage electronic speed controller.

The T200 is a brushless DC thruster capable of around 50N of forward thrust. With slightly higher supplied voltage than specified, the T200 thrusters are running outside their recommended parameters. Brushless DC control methodology uses pulse-width modulation to control the amount of power delivered. In this case, the absolute amount of power delivered to the ESC has been limited to 350W to avoid overpowering the thrusters.

Engineering firm Connett Engineering was contracted to produce prototype azimuth mounts for the T200 thrusters. These were machined out of acetyl material – strong, long-lasting and self-lubricating. An off-the-shelf servo motor was used to actuate each mount shaft.

Figure 7 a and b: T200 thruster (top) and custom-made azimuth mount.

8. PERFORMANCE ANALYSIS

Initial performance analysis was conducted using three prediction analysis algorithms: the Savitsky pre-planing, Savitsky planing and the Holtrop methods. Applying these algorithms provided useful and valuable analysis of the potential performance of this hull form. The application of these algorithms was considered a guide only in analysing the performance of this concept.

The next step was to undertake extensive performance validation of the prototype. While this validation considered the performance of the hull, the major focus was on the whole vessel concept and its performance against the design brief. The measured and observed performance of the vessel has validated the concept.

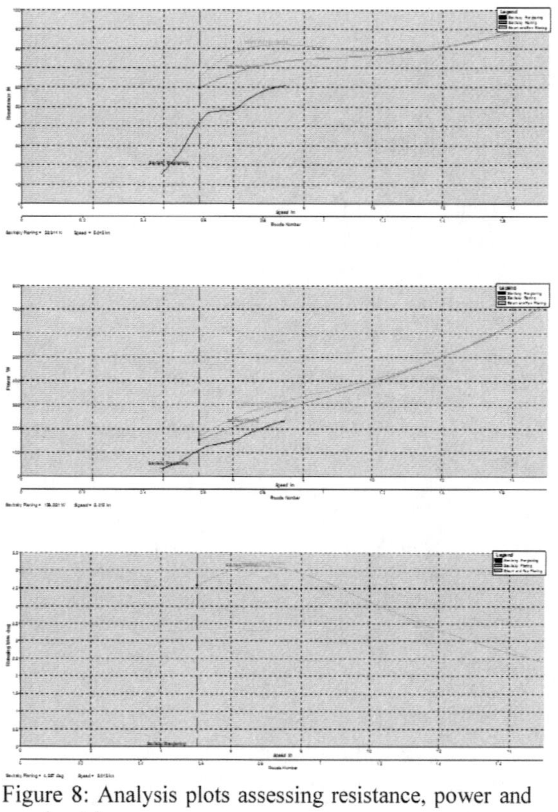

Figure 8: Analysis plots assessing resistance, power and running trim.

7. CONCLUSIONS

The capability to enhance traditional navigational techniques with technology to accurately see what lies beneath the water adds an additional dimension of safety for super and mega yachts wanting to explore inadequately charted areas and harbours and other waterways with changing seafloor conditions. Using an unmanned surface vehicle to enhance exploration capability and improve navigation is a natural step, drawing on advancements in both composite construction and remote operation technology.

The Constellation USV design aims to address the low specification of other USV models and create an efficient hull form which maximises performance under battery power. The interchangeable moon pool module offers flexibility and ease of use. The resulting prototype creates a remote vessel with high-definition survey capabilities that is light, versatile and highly efficient, providing an innovative navigational tool for safe operation of large vessels such as super and mega yachts in less frequented waters.

8. REFERENCES

1. See https://asmt.ae/platforms/hydra/ and Byrne, Diane, 'Hydra USVs for Mega Yachts', https://megayachtnews.com/2015/03/hydra-usvs-for-megayachts/

2. See www.balticyachts.fi/yachts/baltic-197-hetairos/

3. P Dassi, 'Advanced composites: a possible answer to mega yacht structural questions', *Design and Construction of Super and Mega Yachts*, 2015.

9. AUTHOR'S BIOGRAPHY

Robert Shaw heads the New Zealand Maritime School at the Manukau Institute of Technology in Auckland, New Zealand. He is also the principal of performance yacht design studio Shaw Yacht Design, leading collaboration and consultation on custom performance-yacht design projects.

MULTI-CRITERIA OPTIMIZATION AS KEY TO YACHT PERFORMANCE IMPROVEMENT

K Hochirch, S Deucker and **V Bertram**, DNV GL, Germany

SUMMARY

The paper describes the application of multi-criteria optimization to monohull and multihull yacht configurations. The paper outlines the process (involving concept exploration, formal optimization and simulations), key techniques, employed software tools and results. The analyses combine efficient low-fidelity design exploration with high-fidelity CFD simulations for accurate results. The process allows advancing unconventional designs in relatively short time exploiting the power of parallel computing and virtual prototyping. For yachts, "performance" improvement encompasses besides energy efficiency also aspects of passenger comfort. Examples, sometimes anonymized, from industry practice are used to illustrate approach and possible achievements.

1. INTRODUCTION

Traditionally, ship design has been based on experience. Increasingly we rely on "virtual experience" from simulations to guide design decisions, particularly for high-performance vessels such as megayachts. Scope and depth of these simulations have developed very dynamically over the past decade. For hull design, formal optimization for realistic operational profiles has become the industry's best practice, [1].

For megayachts, design goals and constraints differ from cargo vessels. For example, aspects of comfort and habitation play a key role in megayachts. Correspondingly, we see associated analyses for these vessels ranging from simulations of the wind flow around yacht superstructures to seakeeping analyses, [2]. By now, the design process for avant-garde designs relies heavily on virtual prototyping and simulations, leading not only to more efficient designs but also opening the door to innovative and unconventional yacht architectures.

We distinguish then the following design approaches, [3]:

- Simulation-based design
  Simulation-based hull improvement is commonly employed in industry. In this design approach, a few (typically less than 10) hull variants are generated and assessed by computational fluid dynamics (CFD) and/or model tests. This standard design procedure is employed by many design offices and model basins around the world, but frequently we see that significant improvement beyond this approach is possible.

- Concept exploration models (CEMs)
  Concept exploration models (CEMs) have been proposed as an alternative to 'automatic' optimization, [4]. More recently, "design of experiment" (DOE) has been used as a term for the same design strategy: A large set of candidate solutions is generated by varying design variables. Each of these solutions is evaluated by key performance indicators and stored. CEMs thus generate a "map" of the unknown design space. Using suitable graphical displays, the designer gets a feeling how certain variables influence the performance of the design. The approach was deemed impractical for ship design in the 1990s due the then excessive computational requirements. However, parallel computation has changed this, and concept exploration is now used in commercial projects.

- Optimization
  Optimization looks at thousands or even tens of thousands of designs and uses an optimization algorithm to find the best solution. For many modern optimization problems, genetic algorithms (GAs) or related evolutionary optimization algorithms are the preferred choice today. GAs are significantly less efficient than older gradient-based search algorithms. How-ever, they are easily parallelized and robust in finding global optima, i.e. they do not get stuck at local optima. (Single-objective) optimization is in theory a mathematically well-posed problem. However, the objective and all constraints must be expressed as mathematical functions. This is not easy in practice, [4].

- Multi-objective optimization
  Optimization for multiple objectives is strictly speaking nonsense. Mathematically you can only optimize for one objective, respectively an optimum is only defined for one objective function. In layman terms, finding the fastest (objective: speed) and cheapest (objective: price) car will result in the question: Make up your mind, what do you want? Multi-objective optimization in practice is short for a combination of concept exploration and optimization. The concept exploration helps in making up one's mind. Then objectives may be reformulated as constraints or combined in one artificial objective function using weights for the individual objectives. The "best compromise" objective function can then be determined formally by the optimization algorithm of choice.

We will present here recent industry projects, which give insight into how optimization and simulation are applied to develop high-performance designs. The focus of optimization in industry projects lies mostly on reducing required power. While fuel efficiency is still rarely an issue in megayachts, smaller engines mean less weight, less volume, increased range and generally fewer vibration problems.

2. SAMPLE PROJECTS

2.1 YACHT

Blohm & Voss wanted a hull lines optimization in order to improve calm-water performance as much as possible for a given configuration to serve as base for a type class. Such a project can be considered as a typical application of hull lines optimization. Details of appendages may then be redesigned and optimized to improve seakeeping, vibrations or exhaust gas spread.

The process in this project can also be considered as typical, involving a first concept exploration phase which serves to reduce the search space and to elucidate the design objectives and constraints.

The starting point of the optimization was a fully parametric hull description developed within the CAESES - Framework. The baseline design was described using 51 free-form parameters for hull form variation, Fig.1. The form parameters defined, include for example:

- Shape of basic curves (centre plane curve, flat of bottom, transom edge, etc.)
- Bulbous bow shape, length and height
- Fullness of fore- and aftbody
- Waterline entrance angle
- Parallel midbody length
- Bilge shape
- Transom edge height
- ...

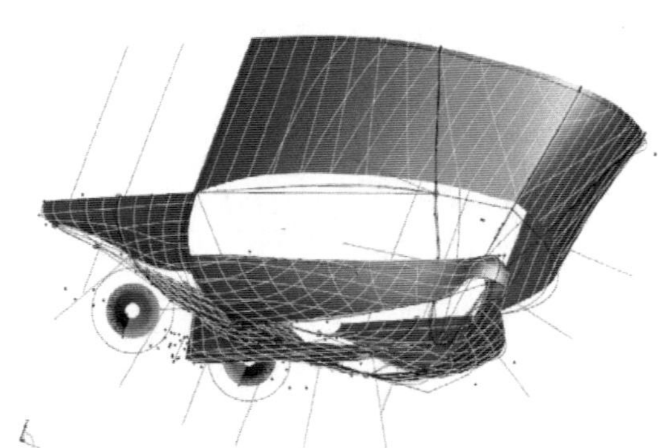

Figure 1: Parametric hull description

Geometrical constraints were specified by the shipyard. These concerned range of the longitudinal centre of buoyancy and displacement, propeller tip clearance, distance from keel to metacentre, bulbous bow tip, and hard points to accommodate main engine and equipment.

The yard also supplied a condensed operational profile, featuring 1 draft and 3 speeds, with associated weights ranging between 10 and 70% which was considered representative for the service of such vessel.

The initial design space exploration explored ~18000 designs, of which ~4300 fulfilled all constraints. It is intended to explore beyond the strict domain of constraints to supply decision support on whether constraints should be softened (if that would result in significant additional improvement). After assessing the results, the objectives and constraints were largely kept. Only the range for longitudinal centre of buoyancy was slightly reduced and 3 more hard points were added in the forebody to assure anchor drop.

A global hull optimization was performed using a fast, simplified hydrodynamic assessment of the generated hull variants. The potential flow solver FS-Flow computed the wave resistance. This fully nonlinear wave resistance code captures the dynamic trim and sinkage iteratively. The viscous resistance was approximated by a semi-empirical friction line, considering the actual wetted hull surface and local Reynolds numbers, thus capturing directly some of the effects of hull changes on the viscous pressure resistance. The propeller thrust was simulated by a propeller body force model. Such a simplified hydrodynamic model allows exploring a vast number of possible designs. But it also requires an indirect and approximate formulation of the objective function, as a direct estimation of the power (and thus fuel consumption) with reasonable accuracy is not feasible.

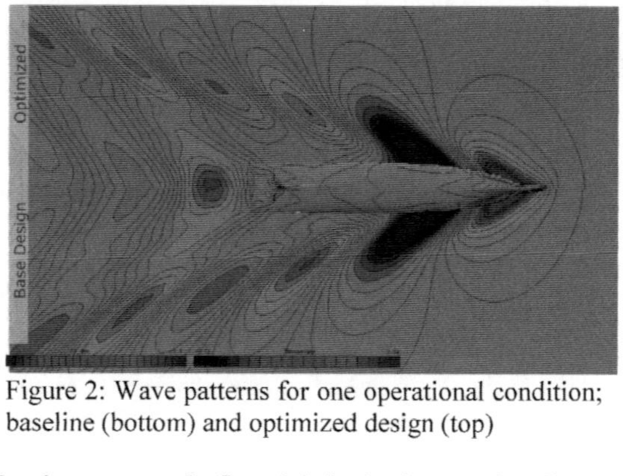

Figure 2: Wave patterns for one operational condition; baseline (bottom) and optimized design (top)

For the purpose of a first global selection, we then formulate several objectives that address partial aspects of good resistance and propulsion:

- Thrust (using empirical estimates for viscous resistance parts and propeller efficiency)
- Total resistance based on wave pattern resistance
- maximum crest-to-trough wave height along the ship hull

- hydrodynamic fairness parameter along the hull

Some 10000 additional designs were generated, of which ~3000 fulfilled all constraints. Fig.2 compares for one operational condition the computed wave patterns of the baseline design and our optimized design. The relatively small changes resulted in 7% improvement for this speed. For the medium speed, improvements of 5% were obtained, accepting a decrease in performance of 4.5% for the lowest (and least often used) speed. Displacement increased by 2% given welcome extra margins for the design.

As a final stage, numerical towing tank tests were performed with a high-fidelity CFD (Computational Fluid Dynamics) code, solving the Reynolds-averaged Navier-Stokes equations (RANSE), thus capturing viscous effects directly in the simulation, Fig.3. Based on these results, standard model basin extrapolation procedures gave the predictions for full scale. These high-fidelity predictions gave an improvement of 5.5%.

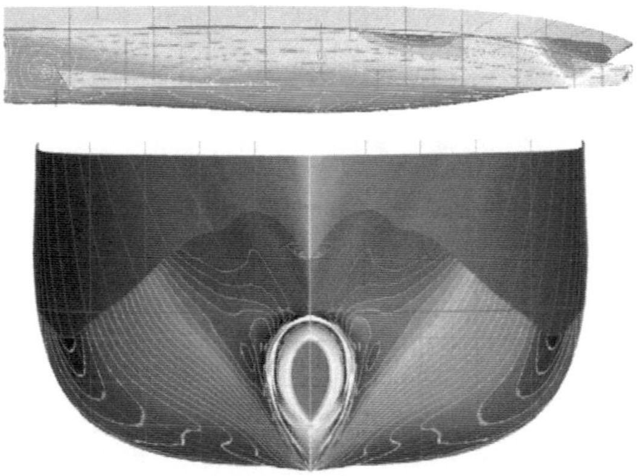

Figure 3: Pressure distribution and wave profile obtained by CFD for one speed

2.2 TRIMARAN

In a recent project with Frers Design for an avantgarde power trimaran yacht, Fig.4, both calm-water performance and seakeeping loads had to be considered, [5]. Again, a concept exploration with hydrodynamic assessment based on a nonlinear wave resistance considered a condensed matrix of 6 draft-speed combinations reflecting the operational profile.

Figure 4: Design of the power trimaran

The parametric model describes central hull, outriggers and relative positioning. The concept exploration looked at ~10000 designs. After the concept exploration, a selected number of designs was investigated in numerical towing tank tests using high-fidelity free-surface RANSE simulations, Fig.5. Extrapolation to full-scale sea trial conditions followed standard model basin practice. Model tests at the SVA Potsdam verified the results of the simulation for the resultant hull shape of a second refined design loop.

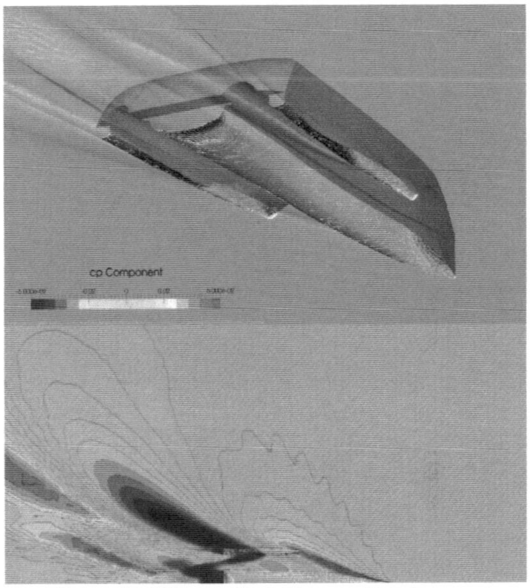

Figure 5: CFD results for 1 operational case

Design loads (vertical bending moment and shear force) were calculated using direct simulations to comply with Class Rules. Due to the unconventional hull shape and the high vessel speeds, standard empirical formulae for roll damping were inappropriate. Instead, RANSE simulations were conducted to determine speed-dependent roll damping coefficients. The damping coefficients determined in these numerical simulations were then used in the 3D potential flow seakeeping code RANKINE to determine short-term accelerations (for safety and comfort) and long-term loads (for structural design). Again, model tests at Potsdam Model Basin (SVA) confirmed the simulations for motions and accelerations for selected cases.

3. OUTLOOK

Comfort aspects have been so far mainly addressed by simulation-based design, where after each simulation human insight is used to derive design changes and only 1 or 2 loops are afforded before the design is accepted. Research applications showing formal optimization of yacht funnel designs, [2], or OSV seakeeping, [6], indicate the [near] future of industrial design projects employing formal optimization also for comfort aspects in megayacht design.

Fig.6 shows exemplarily how motion sickness incidence (MSI) is already considered as an additional optimization objective in industry projects. The application shown is for a ferry, but could be applied similarly for megayachts or cruise vessels.

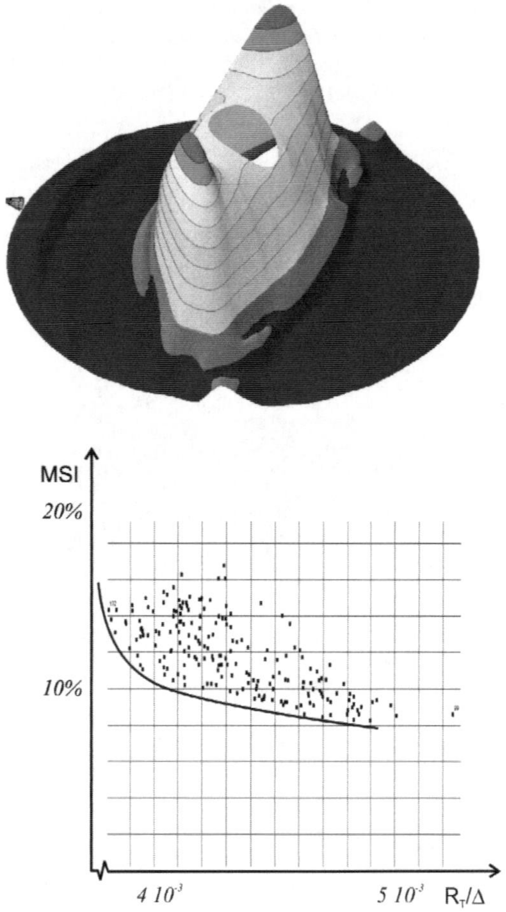

Figure 6: MSI as optimization objective

4. CONCLUSIONS

Experience-based design has been largely replaced by simulation-based design. Currently, we see simulation-based design being replaced by concept exploration and formal optimization.

The megayachts, comfort aspects play a key role in design. In the recent past, they have been addressed mainly by simulation-based design, but are expected to drift into formal optimization as industry best practice over the next decade.

5. ACKNOWLEDGEMENTS

We would like to thank our teams in Potsdam and Hamburg for their support and discussions in writing this paper.

6. REFERENCES

1. HOCHKIRCH, H., BERTRAM, V., 'Hull optimization for fuel efficiency – Past, present and future', *11th Conf. Computer and IT Applications in the Maritime Industries (COMPIT)*, Liege, 2012, pp.39-49.

2. KÖHLMOOS, A., BERTRAM, V., 'Advanced simulations for high-performance megayachts', *8th Int. Conf. High-Performance Marine Vehicles (HIPER)*, Mülheim, 2012, pp.38-50.

3. HOCHKIRCH, H., BERTRAM, V., 'Optimization for Ship Hulls – Design, Refit and Operation', *6th Int. Conf. Computational Methods in Marine Engineering (MARINE)*, Rome, 2015, pp.210-217.

4. BERTRAM, V., 'Optimization in ship design', *OPTIMISTIC – Optimization in Marine Design*, Mensch & Buch Verlag, Berlin, 2003, pp.27-53.

5. BERTRAM, V., KREBBER, B., HOCHKIRCH, K., FRERS, G., FLYNN, M., 'Advanced Simulation-Based Design for Avant-garde Power Trimaran', *11th HSMV Conf.*, Naples, 2017

6. HARRIES, S., LAND, J., WEICKGENANNT, S., QIU, W., 'From Analysis towards Design in Seakeeping', *11th Conf. Computer and IT Applications in the Maritime Industries (COMPIT)*, 2012, Liege, pp.27-38.

6. AUTHORS BIOGRAPHY

Karsten Hochkirch is heading the Fluid Engineering department of DNV GL. The specialty of his group, formal parametric optimization, has been successfully applied to a wide range of ship designs and help to save many thousand tons of CO_2 each day.

Stefan Deucker is Regional Manager Maritime Advisory - Germany at DNV GL. In this role, he is responsible for combining engineering analyses with consulting services.

Volker Bertram is Senior Project Manager at DNV GL. He has extensively worked on CFD and ship design methods and authored two books on these topics.

NEW WAY TO DETERMINE THE NAVIGATION AREA AND RESTRICTIONS ON WEATHER CONDITIONS FOR HIGH-SPEED MEGA YACHTS

O.O. Kanifolskyi, Odessa National Maritime University, Ukraine

SUMMARY

Several different versions of the Rules governing the navigation areas for small vessels currently exist. For example, various categories of vessels are proposed in International Standards ISO, European Union Directives and the Rules of various classification societies (GL, BV, RINA, Shipping Register of Ukraine and others). These regulatory documents contain various names of navigation areas and various values of permissible wave height characterizing the category of the vessel. Some Rules offer formulas for calculating the maximum height of a significant wave that a ship can meet in a voyage. The need to compare the existing requirements of various Rules, as well as to propose a new way of calculating permissible wave heights, which will determine the effective operation of the vessel, are actual tasks. The new way of calculation will be based on a comparison of the total energy of the wave and the kinetic energy of the vessel.

NOMENCLATURE

V	Displacement volume (m^3)
$Fr_V = \frac{v}{\sqrt{g\sqrt[3]{V}}}$	Volume Froude number
$Fr_L = \frac{v}{\sqrt{gL}}$	Froude number
v	Ship speed (m s^{-1})
L	Length of the ship (m)
D	Depth of the ship (m)
B	Width of the ship (m)
d	Draft of the ship (m)
c_b	Block coefficient
γ	Density of water (t m^{-3})
n	Overload
a_{CG}	Design vertical acceleration at LCG
h_w	Height of the wave
L_w	Length of the wave

1. INTRODUCTION

When assigning navigation area or weather restrictions for a ship, engineers usually follow the Rules of classification societies and take into account many parameters: strength and hull design, mechanical installation, stability, unsinkability, ship facilities, and the technical condition of the ship.

When assigning a navigation area and restrictions on weather conditions for high-speed vessels that perform various important tasks and yachts whose owners would like to get the maximum possible use of their own vessel, it is especially important to take into account the possibility of efficient movement at a certain wave height and ensure minimum overload values. Overload n is the ratio of the vertical non-gravitational nature accelerations in the cross section of the vessel, caused by forces, when meeting a wave, to the acceleration of gravity.

Currently, there are many variants for assigning navigation areas for a vessel and restrictions on weather conditions. These ways are offered in various international documents and rules of classification societies. Further, in this article, an attempt will be made to propose a method for determining weather conditions under which efficient operation of the vessel is possible.

It is necessary to compare the kinetic energy of the vessel and the energy of the oncoming sea wave. If the wave energy exceeds the kinetic energy of the ship, the ship will not be able to move forward effectively, in some cases fully losing speed. The steersman will be forced to reduce the speed of the vessel, change course or go around a geographical area. If the kinetic energy of the vessel dominates over the wave energy, the vessel will be able to move forward with some loss of speed.

Different classification societies set overload limits for high-speed passenger and cargo ships. Data on permissible overloads are generally similar, confirmed by experimental studies and can be applied to large yachts. The overload values are mainly related to the relative speed and height of the wave.

2. REVIEW OF THE REQUIREMENTS OF SOME INTERNATIONAL DOCUMENTS AND RULES OF CLASSIFICATION SOCIETIES

What is Speed Mega Yacht? The SOLAS formula used to determine the speed of a high-speed vessel has the form $v_{max} \geq 3.7V^{0.1667}$. Mode of movement of the vessel is characterized by the volume Froude number $Fr_V = \frac{v}{\sqrt{g\sqrt[3]{V}}}$. After transformations, this formula will take the form $v = 3.13Fr_V V^{0.1667}$. After equating the right parts of the conventional inequality and this equation, the possibility appears to determine the volume Froude number that characterizes the speed vessel $Fr_V \geq$ 1.18. The formula linking the relative speeds in the interval of transitional mode, $1 \leq Fr_V \leq 3$, has the view $Fr_L = 0{,}29Fr_V + 0{,}11$ [1]. It follows that a high-speed vessel is a vessel with relative speed $Fr_L \geq 0.45$.

There are no clear definitions for super-, mega- yacht, in the Rules of classification societies. A variety of conventional and national rules often apply to ships longer than 24 meters or 500 GT.

For example, The Large Commercial Yacht Code [2] applies to motor or sailing vessels of 24 metres in load line length and over or, if built before 21 July 1968, which are of 150 tons gross tonnage and over and which, at the time, is in commercial use for sport or pleasure and which carries no cargo and no more than 12 passengers. Applying a simplified IMO method to calculate gross tonnage (GT) [3], we can draw the following conclusions. According to this method, the gross capacity of the vessel is determined by the formula $GT = a V_E$. The coefficient a depends on the value of the cubic module $V_E = LBD$ and is selected from the corresponding table. For a vessel with a cubic module less than 400, the coefficient a is set to 0.58. Substituting in the formula the average relative values that are characteristic for high-speed motor yachts with a relative speed $1 \leq Fr_V \leq 3$, $\frac{B}{d} = 4.7$ (NPL data, minimum resistance of water), $\frac{L}{B} = 6$ (well-flowed body) and the ratio $\frac{D}{d} = 1.5$ (mean value typical for speed megayacht, connected with requirements of the Convention on Load Lines and unsinkability), we obtain the value of the length of the ship corresponding to different values of GT, Figure 1.

GT	150	300	400	500
Ship length, m	31	39	43	46

Figure 1. The dependency of the GT on ship length.

So, for this study selected yachts of length equal or more than 24 m.

The areas and speeds of operation of high-speed vessels are limited by the allowable design acceleration $a_{CG} = g$ (value for high-speed passenger and cargo ships) [5] and the corresponding overload value $n = \frac{g+g}{g}$. For the calculations, a fixed value n = 2 was adopted, recommendations by Bureau Veritas (for passenger ships n= 2, for pleasure boats n= 2.3) [4]. Similar values are proposed in the rules of Germanischer Lloyd (GL) [5], for relative speed $Fr_V = 2$, midpoint of the transitional mode. The value of the coefficient n changes depending on the relative speed [5].

Basically, all rules determine the navigation area of vessels and their design category, depending on the height of the wave and the possible distance from the harbour. In this work, we will consider the limitations of the navigation areas associated with the height of the waves. Rules [5] define the navigation areas listed in the figure 2. The Shipping Register of Ukraine offers the following areas [6], figure 3.

Sea areas	Wave height $h_{1/3}$	Wave height $h_{3\%}$
"Open-sea service"	$h_{1/3} \geq 4$ m	$h_{3\%} > 5{,}3$ m
"Restricted open-sea service"	$2{,}5\ m \leq h_{1/3} < 4$ m	$3{,}3\ m \leq h_{3\%} < 5{,}3$ m
"Moderate environment service"	$0{,}5\ m < h_{1/3} < 2{,}5$ m	$0{,}7\ m < h_{3\%} < 3{,}3$ m
"Smooth sea service"	$h_{1/3} \leq 0{,}5$ m	$h_{3\%} \leq 0{,}67$ m

Figure 2. The sea areas, GL.

Navigation areas	Wave height $h_{3\%}$
R1	8.5 m
R2	7 m
R2-RSN	6 m
R3-RSN	3.5 m
R3	about 2 m
Berth-connected ship	

Figure 3. The navigation areas, Shipping Register of Ukraine.

In Rules GL, the term "Significant wave height" $h_{1/3}$ is used. "Significant wave height" is the average height of the one third highest observed wave heights over a given period. In other Rules, for example, the Shipping Register of Ukraine uses height $h_{3\%}$ (height with 3 percent probability). A comparative analysis for these determinations of wave heights is given on the figure 2, $h_{3\%} = 1.33h_{1/3}$.

The Directive 2013/53/EU on recreational craft and personal watercraft defines the following design category, figure 4 [7].

Design category	Wind force (Beaufort scale)	Significant wave height $h_{1/3}$, m
A	exceeding 8	exceeding 4
B	up to, and including, 8	up to, and including, 4
C	up to, and including, 6	up to, and including, 2
D	up to, and including, 4	occasional waves of 0,5 m maximum height

Figure 4. The design category, Directive 2013/53/EU.

There is a standard ISO [8]. This standard is principally applicable to boats propelled by human or mechanical power of 6 m up to 24 m hull length. The following categories are defined by the standard figure 5.

Design category	A	B	C	D
wave height up to	approx. 7 m significant	4 m significant	2 m significant	0.5 m maximum
Typical Beaufort wind force	up to 10	up to 8	up to 6	up to 4

Figure 5. The design category, ISO.

Another well-known document "LY3" [2] defines the following navigation areas. In general, requirements given within the Code are based on unrestricted geographical operation outside Polar Regions, however, where considered appropriate, standards for vessels operating as Short Range Yachts have been included.

"Short Range Yacht" means a vessel under 500GT the keel of which was laid or was at a similar stage of construction prior to the 1st August 2005 or a new vessel constructed on or after that date under 300GT: restricted to operating in forecast or actual wind of a maximum Beaufort Force 4, for a motor yacht, and Beaufort Force 6 for a sailing yacht.

The Shipping Register of Ukraine offers restrictions on weather conditions. Non-passenger ships with a length of 20 to 24 m - $h_{1/3}$= 3.3 m. Passenger ships of length between 20 and 24 m - $h_{1/3}$= 1.4 m.

Some rules propose to determine the maximum value of the wave height with which a ship can meet. For example, the formula $h_{m1/3} = 5\frac{a_{CG}}{v/\sqrt{L}}\frac{L}{6+0.14L}$ [5], includes: length of the vessel, speed and design acceleration, the values of which should not be exceed $1g$, for passenger and cargo vessels. In this case, the overload value is $n = 2$.

3. METHOD OF DEFINITION OF THE NAVIGATION AREA AND RESTRICTIONS ON WEATHER CONDITIONS FOR SHIPS.

Energy $E_W = \frac{\gamma g h_W^2 L_W}{8}$ characterizes energy per unit width of the wave surface, height h_W and length L_W. The relation between the parameters of the wave is assumed to be equal to $\frac{L_W}{h_W}$=20. Next we find the kinetic energy of the vessel $E_S = \frac{1.1mv^2}{2} = \frac{1.1\gamma c_b LBdv^2}{2}$, taking into account the attached mass of water.

The specific energy of the wave, for all width of the ship, after the division of the equations for the vessel and wave energies to the draft, has the form $\bar{E}_W = \frac{\gamma g h_W^2 L_W}{8}\frac{B}{d}$. The specific energy of the ship is $\bar{E}_S = \frac{1.1\gamma c_b LBdv^2}{2}\frac{1}{d}$.

Then write inequality, which compares the specific kinetic energy of the vessel and the specific energy of the wave $\frac{1.1\gamma c_b LBdv^2}{2}\frac{1}{d} \le \frac{\gamma g h_W^2 L_W}{8}\frac{B}{d}$.

After the transformations we get the following inequality

$$\frac{1.1c_b LBv^2}{2} \le \frac{g h_W^2 L_W}{8}\frac{B}{d}. \qquad (1)$$

The relative length $\frac{L}{B}$ and relative width $\frac{B}{d}$ of the vessel were chosen, as described above. The value of the block coefficient is c_b= 0.5 (minimum resistance of the water in the range of relative speeds $1 \le Fr_V \le 3$). We can get the length of the ship. So, it is possible to calculate the length of the vessel having a kinetic energy equal to the wave energy, inequality (1).

The wave height $h_{3\%}$= 3,5 m ($h_{1/3} = 2.6$ m) corresponds to the navigation area R3-RSN, the "Restricted open-sea service" and the "B" category. The results of calculations for inequality (1) are presented on the figure 6.

v, m/s	5	10	15
L, m	66	33	22
Fr_L	0.2	0.56	1.02

Figure 6. The results of calculations for inequality (1), $h_{1/3} = 2.6$ m.

The data from the table is shown on the figure 7.

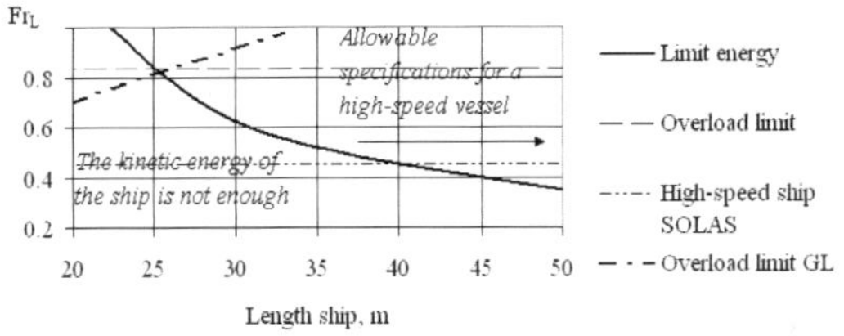

Figure 7. Graph for assigning wave height restrictions for a ship, $h_{1/3} = 2.6$ m.

Analyzing the graph it can be noted that, when the wave height $h_{3\%}$ = 3.5 m. ($h_{1/3}$ = 2.6 m), high-speed vessel has sufficient kinetic energy when it has a length of 38 m and a relative speed Fr_L= 0.45. Such a vessel will be able to move at a reduced speed. For large lengths and relative velocities, the energy of a high-speed vessel also exceeds the energy of the wave. Ships having a length of less than 38 m should have a higher relative speed and, accordingly, a greater value of kinetic energy, in order to overcome sea waves. Otherwise, during the movement of such vessels at oncoming sea waves, there will be a need to change course, reduce speed, to drift and effective operation of the vessel will become impossible. At the top of the graph, two more lines are plotted. The dash-dotted line shows the limit of overloads calculated using the formula $h_{m1/3} = 5\frac{a_{CG}}{v/\sqrt{L}}\frac{L}{6+0.14L}$ [5], for a given wave height. The calculation adopted the upper permissible limit of design acceleration $a_{CG} = g$ and, accordingly, the permissible limit of overload $n = 2$. The second dashed line was obtained on the basis of data from the work of [9] on the connection of overload, relative velocity and wave height. It follows from the graph, that the smallest length of a vessel with sufficient kinetic energy and acceptable design acceleration values is 26 m, for $h_{1/3} = 2.6$ m.

For further development of this idea, in relation to large yachts, a calculation was made for wave height $h_{3\%}$=5.3m., $h_{1/3}$ = 4 m, figure 8.

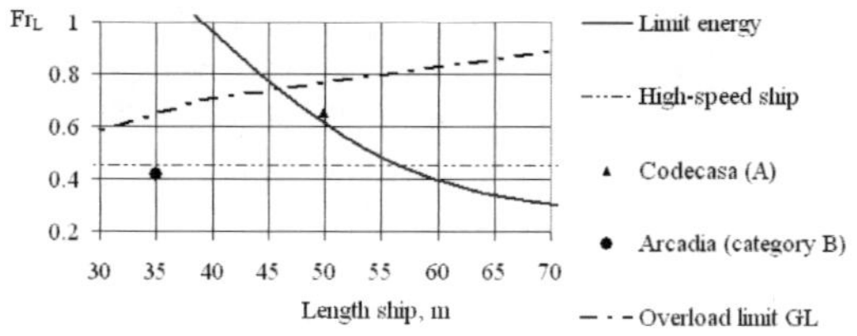

Figure 8. Graph for assigning wave height restrictions for a ship, $h_{1/3}$ = 4 m.

The figure shows that a high-speed vessel, with sufficient kinetic energy, for a wave with a height of $h_{3\%}$ = 5.3 m, ($h_{1/3}$ = 4 m), has a length of 57 m and a relative speed Fr_L = 0.45. From the graph it follows that the smallest length of the vessel with sufficient kinetic energy and acceptable overload values is 46 m. In the figure, points characteristic of large yachts with different design categories are additionally plotted.

4. CONCLUSIONS

The method of solving the task of assigning a navigation area and restrictions on weather conditions for high-speed large yachts is proposed. In determining these criteria, calculations based on data on the total energy of the wave, kinetic energy and relative speed of the vessel, and taking into account the restrictions of classification societies, with regard to overloads, must be made.

If we consider the limit on weather conditions with a wave height of $h_{3\%}$= 3.5 m, $h_{1/3}$= 2.6 m, then the minimum length of the effective vessel (sufficient kinetic energy and acceptable overload) will start from the value L= 26 m, with the corresponding Froude numbers.

The acceptable length of an effective high-speed vessel (sufficient kinetic energy and permissible overload values), at a wave height of $h_{3\%}$= 5.3 m, $h_{1/3}$= 4 m, will depend on its relative velocity and will start from a value of about 46 m. This length value is in good agreement with the gross tonnage (GT500 and length, see the Figure 1) of Short Range Yachts [2]. For ships of this category, it becomes possible to increase their navigation area to wave heights $h_{3\%}\leq$ 5.3 m, $h_{1/3}\leq$4 m, instead of 6 Beaufort points $h_{3\%}\leq$ 3.5 m, $h_{1/3}\leq$ 2.6 m.

5. REFERENCES

1. KANIFOLSKYI O.O., 'Definition of main dimensions of high-speed displacing monohull ships on the first steps of project', *Odessa National Maritime University*, 2003.
2. MARITIME AND COASTGUARD AGENCY, 'LY3. The Large Commercial Yacht Code', *Maritime and Coastguard Agency*, 2012.
3. MARITIME SAFETY COMMITTEE, 'Simplified tonnage calculation for existing ships', *International Maritime Organization*, 1994.
4. BUREU VERITAS, 'Rules and Regulations for the Classification of Ships of less than 65 m in length. Part II-B. Hull Structure. Chapter 13. Light highspeed ships', *Imprimerie strasbourgeoise*, 1990.
5. GERMANISCHER LLOYD, 'Rules for Classification and Construction. Chapter 5. High Speed Craft', *Gebrüder Braasch, Humburg*, 1996.
6. SHIPPING REGISTER OF UKRAINE, 'Rules for Classification and Construction of Ships', *Shipping Register of Ukraine,* 2010.
7. EUROPEAN PARLIAMENT AND COUNCIL, 'Directive 2013/53/EU on recreational craft and personal watercraft', https://eur-lex.europa.eu/eli/dir/2013/53/oj, 2013.
8. ISO, 'Small craft — Stability and buoyancy assessment and categorization - Part 1: Non-sailing boats of hull length greater than or equal to 6 m', *ISO*, 2002.
9. LOKSHIN A.Z., 'Design of ship hull. Hydrofoils', *Leningrad Shipbuilding Institute*, 1978.

6. AUTHORS BIOGRAPHY

Oleksandr Kanifolskyi holds the current position of Docent of the chair "Theory and Designing of ship" at Odessa National Maritime University. He is responsible for teaching the courses of lectures about International Maritime Conventions, Basic Ship Theory, Ship Hull Construction, etc. Together with colleagues, he has work about improvement construction, strength and seagoing of ships and increasing to efficiency of the functioning the fleet and ports. His experience includes the publication of several articles on the design of small high-speed transitional mode vessels.

DESIGN AND SIZING OF BATTERY SYSTEM FOR ELECTRIC YACHT AND FERRY

U Malla, Hornblower Cruises and Events, USA

SUMMARY

The need for Zero Emission and clean energy vessels has been a major importance in the recent years. The aim of the paper is to design and size the battery system for Luxury yachts and Ferries using the Li-ion battery technology. The evolution of the battery technology in the last decade made the yacht / ferry owners to implement the battery technology to the electric vessels. This paper discusses the step by step approach for the battery system arrangement, sizing of the battery racks, selection of the cooling systems for batteries, effect on stability, cost, safety and design life of the batteries.

In this paper it summarizes the cost benefit analysis of the battery system for the electric yacht / ferry when compared to a conventional diesel / hybrid propulsion system.

NOMENCLATURE

ESS	Energy Storage System (kWhr)
DoD	Depth of Discharge (%)
D	Design Life (years)
N	No. of trips per year
CAPEX	Capital Expenditure
OPEX	Operating Expenditure
Li-ion	Lithium Ion Battery Technology
$	US Dollar
BMS	Battery Management System

1. INTRODUCTION

Increasing awareness of climate change is emerging and the marine transport sector is exposed to more demanding regulations to reduce impacts. Research studies conducted by companies with commercial interests in the ferry market, confirm that all-electric car ferries have less impacts on global warming than conventional ferries driven on marine diesel oil. The challenge with most of these studies is the lack of clear methodology or inaccessible data and calculations.

Worldwide, there are more than 20 million pleasure boats and nearly 10 million of them are in the USA. Approximately 236,000 metric tons of hydrocarbons from boats finds it's way into the water every year, equivalent to more than seven times the amount spilled by the Exxon Valdez in Alaska in 1989. Many boat motors are of the 2-stroke design that burn a mix of oil and gas. These types of motors may be eight times more pollutant than four-stroke motors, since up to 25% of the fuel might be emitted half-burned through the exhaust into the water.

The marine shipping industry is one of the fast growing industry which account for 3% of the world CO2 gas emissions to the environment and is expected to grow to 5% by 2050. All the maritime governing bodies such as governments, ports and marina across the world are pushing towards the zero emissions for the marine vessels.

This paper describes the steps to design and size the electric vessel using the in-house Hornblower vessel data and research. It also demonstrates the comparison of CAPEX and OPEX costs of lithium-ion batteries and electricity in the US and the costs associated with typical marine diesel engine for ferries / yachts

3. EVOLUTION OF BATTERY TECHNOLOGY

The evolution of battery technology moving from lead-acid to lithium-ion with high storage capacity to size ratio and low cost to storage. The cost of Li-Ion cells for cars dropped from some $500/kWh five years ago, to $250/kWh and is expected to continue to decline to less than $100/kWh by 2020. The energy density of Li-Ion cells improved from 200Wh/kg 10 years ago to approximately 270Wh/kg today.

The revolution will not stop with the car industry. Niche markets such as ships, boats, yachts, watercraft, drones and underwater vehicles will eventually benefit from the reduction in battery price.

Li-ion is already offering these markets batteries with four times more energy than lead-acid and a 10-year life cycle.

The initial cost of adopting Li-Ion batteries is presently higher than of lead-acid but will become more attractive in the near future. While lead acid will not disappear so quickly, niche markets where volume, weight and logistics are important will lead the way towards the adoption of lithium batteries.

Great strides have been taken to address safety issues related to Li-Ion. Battery makers are now incorporating fire suppressive materials and technologies into their batteries to facilitate fast incorporation of lithium batteries.

In recent years, progress has been made with the development of battery powered ships using li-ion batteries. The first all electric Norwegian ferry MF Ampere undertakes 56 journeys of 5.6 km per day, and is powered by 1.04 MWh Lithium-ion battery which is recharged by two 410 kWh shore charging stations located at the shore end of the journey .Similarly, the recent vessel "Future of the Fjords" launched in the year 2018 with a battery storage capacity of 1.8 MWh awarded the "Ship of the Year 2018" for its innovation and zero emission vessel.

4. OVERVIEW OF ALL ELECTRIC VESSELS

The all electric vessel has the primary power from the battery which serves the propulsion unit and utilities (i.e. electric loads for accommodation). There are three electric propulsion system configurations which are Diesel Mechanical type, Diesel Electric type and all electric type. Considering same propeller efficiency for all the three propulsion systems, the most efficient is all electric system with 68% efficiency when compared to Diesel Mechanical is 32% and Diesel Electric is 28%.

The all electric vessel configuration consists of Battery Racks, Variable speed drive, propulsion motor and propeller as shown in Figure [1].

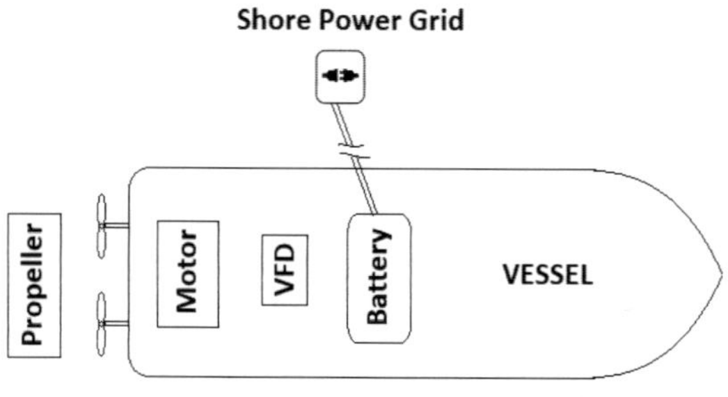

Figure 1: All Electric Vessel System Configuration

A battery powered ship would have no need for fuel tanks, fuel processing, exhaust and air trunking, the diesel engine and its gear box. Current batteries are limited in lifecycle (maximum is 10 years but typically 5 years) which means they may have to be replaced several times throughout the ship lifetime. The battery replacement cost together with the cost of electricity consumed from the shore power grid are the major operating cost of the propulsion plant but this is offset by the fact that there is no longer any need for bunkering diesel fuel and regular maintenance of diesel engines. Electric systems tend to be reliable and easily reconfigurable so an unmanned vessel might be the way for future marine transportation.

5. BATTERY SYSTEM DESIGN AND SIZING

The all electric vessel is designed by evaluating the following steps as shown in flowchart Figure [2]:

1. Owner Requirements
2. Sizing of Battery
3. Battery System Arrangement
4. HVAC Design
5. Fire and Safety Design
6. Vessel Design Checks
7. Cost Analysis

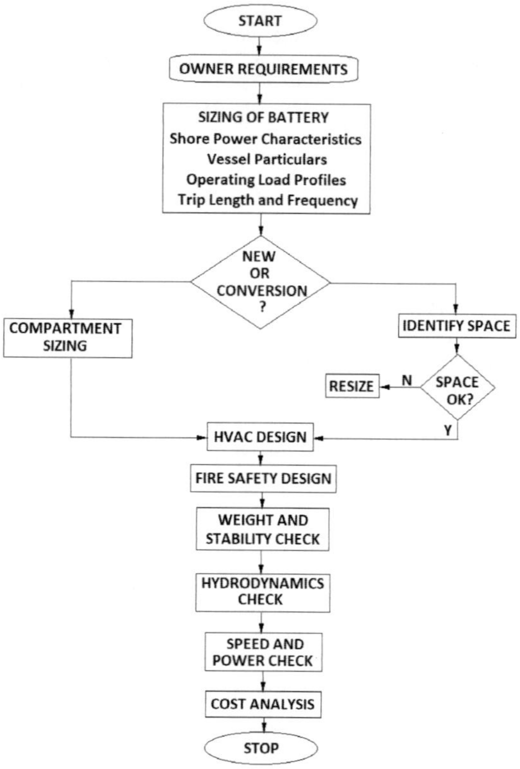

Figure 2: Flowchart for the Design and Sizing of Battery System

In the present paper we consider Lithium ion battery technology as the prime power generation for the vessel design and sizing.

4.1 OWNER REQUIREMENTS

The design of any marine vessel starts with the gathering of requirements from the owner which are typically the speed of the vessel and deadweight i.e. type of cargo / passengers on the vessel. For the all electric vessel the speed of the vessel, carrying capacity of the vessel, the utility needs, length of the journey and frequency in order to determine the operational loads for the sizing the battery racks.

4.2 BATTERY SIZING

The Li-ion cell has a voltage of approximately 3-4 V and several cells are connected together in a module to obtain larger voltage. A group of modules are coupled together to obtain larger voltage and capacity which are called as battery rack or string. The Figure [3] delineates the components of a battery rack.

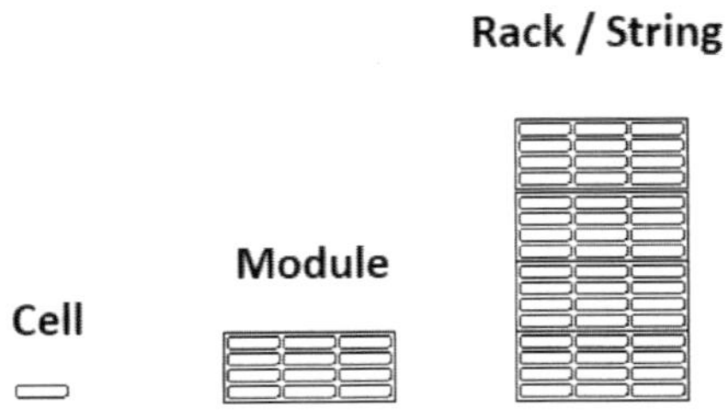

Figure 3: Battery Package Definition

The battery sizing for the vessel depends of the following factors:

- Shore Power Characteristics: Based on the voltage and current limitations on the shore the size and the number of racks of the battery is limited. The height of the battery rack is also dictated by the clear height of the battery compartment with certain allowance for tool access
- Vessel Particulars: The width and depth of the vessel sometimes dictates the arrangement and height of the battery racks in order to attain effective stowage of the batteries in the battery compartment.
- Operating Load Profiles: The battery energy storage system depends primarily on the loads from the propulsion systems, utilities and auxiliary machinery. By plotting a graph between the power load and the time during the journey for the operating profile of the vessel as shown in Figure 4.

Figure 4: Operations Loads Profile

- Trip Length and Frequency: The size of the battery storage capacity depends on the length of the trip journey and the frequency as this affects the design life predominantly.

The limitations of the battery system will greatly dependent of the Heat generation, material degradation and hardware limitations. The energy demand, depth of discharge cycles and the temperatures of the battery compartment are the major factors which dictate the battery system size.

Battery systems will have a specified maximum charging- and discharging current. Ohm's law dictates that the heat generated by the battery cells will be proportional to the current squared, and heat is a major contributor to degradation of the battery cells. Working continuously with relatively high currents will therefore greatly affect the lifetime of the battery cells, and one will therefore have to increase the battery capacity in order to decrease the relative current on each battery cell to conserve lifetime. Moreover, energy movement will also cause capacity fade, and the deeper these discharge cycles are, the lifetime will be shortened exponentially. The below Figure [6] shows a typical Li-ion Battery life cycle vs depth of discharge.

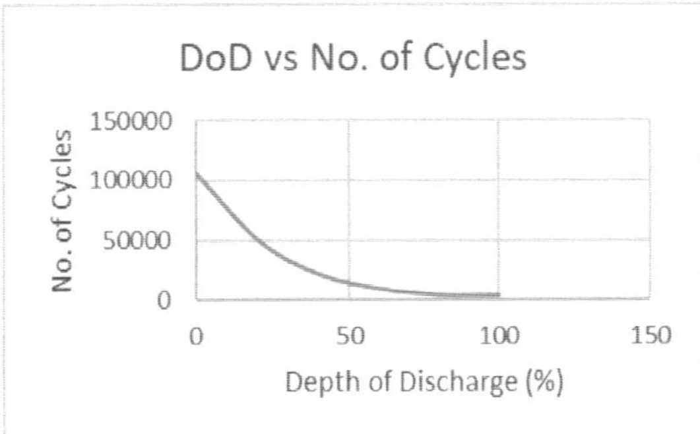

Figure 5: Depth of Discharge vs No. of Cycles

The following is an empirical formula arrived based on the past electric vessel data, lithium-ion battery properties and first principles of engineering in order to arrive at the battery storage size at first cut for preliminary estimation of cost at business front.

$$ESS = \frac{400 \times Load}{[332 - 28.98 \ln(N \times D \times 360)]}$$

Where,

ESS - Energy Storage System kWhr
N - No. of Trips per day
D - Design Life in years
Load - Total Load on the battery in kWhr

4.3 BATTERY SYSTEM ARRANGEMENT

Once the battery sizing is done the arrangement of the battery racks in the battery compartment by considering the following aspects:

- Location of vents
- Weight distribution
- Escape route plan
- Walkway platform
- Battery Replacement Plan
- Size of the Battery Racks

4.4 HVAC DESIGN

The design life of the battery depends mainly on the temperature of the battery compartment. The battery compartment needs to be maintained at the range of design temperatures in order to maximise the design life. Therefore, the HVAC design of the battery compartment needs to done with proper ventilation, heating / cooling and insulation.

The Heating and cooling loads in the battery compartment needs to be calculated by considering the following components:

- Heat dissipated by batteries
- Heat transfer from side shell of the vessel
- Heat transfer from vessel bottom and main deck
- Load due to lighting.
- Infiltration Loads
- Loads from fan motors and BMS.

The capacity of the cooling / heating equipment is determined based on the corresponding cooling and heating loads along with the ducting design.

4.5 DESIGN LIFE

The design life of the battery system depends up many factors i.e. heat generated by the batteries, depth of discharge cycles and current in the batteries. The Li-ion batteries have a typical 10,000 cycles at 35% DoD. The design life of the battery dictates the breakeven period for the all electric vessel when compared to a diesel mechanic engine. Most Li-ion batteries are typically designed for a design life of 10 years.

4.6 FIRE SAFETY DESIGN

The battery compartment with the battery racks needs to be designed with Fire suppression system as per the DNV / ABS guidelines for Li-ion battery installations.

The structural boundary of the battery compartment should provide adequate protection to other parts of the vessel in the event of a battery fire. Fire integrity should be A-60 in the overhead and towards any machinery spaces, passenger/crew spaces, or fuel tanks. The compartment should be provided will the fire safety items as per the SOLAS / IMO.

The battery equipment needs to be designed with sufficient fire protection and safety as per manufacturers recommendation.

4.7 VESSEL DESIGN CHECKS

The electric vessel design with the finalised battery arrangement the following design aspects needs to be checked:

- Weight and Stability: The mass properties of the battery systems from the battery vendor are used to calculate the light weight and the loading condition of the vessel. The stability of the vessel is checked to meet the IMO / Class guidelines.
- Hydrodynamics: With the vessel positive stability results, the vessel is checked for the hydrodynamic analysis to determine the seakeeping and manoeuvring characteristics of the vessel to be within the design limits.
- Speed and Power: The speed and power requirements of the vessel needs to calculate with great accuracy when the vessel is a conversion from diesel engine to electric vessel. The trade off between the weight of the added battery systems and the fuel tank removal dictate the new speed / power of the vessel.

5. ECONOMIC FACTORS VIABILITY

The major decision factor for selecting the electric vessel is the cost comparison between a traditional diesel 2 or 4 stroke engine and battery system operated vessel. The CAPEX for the electric vessel is typically around three times when compared to a 2 or 4 stroke engine but the OPEX is almost 40% lower with the advantages of having zero emissions to the environment and no noise / vibrations.

With the advancement of the Li-ion battery technology, reduced costs of the battery, increased safety and increased design life the viability of the electric vessel is getting more prominent. Based on the estimated calculations for a 2.4 MWhr battery Energy storage capacity vessel with the associated CAPEX and OPEX the payback period will be approximately 4 years with the current prices for the battery in the market.

6. CONCLUSIONS

This paper demonstrates the step by step approach for the design and sizing of the battery system for the "Zero Emission" all electric vessels. It also discusses about the various design aspects of the electric vessel design along the cost comparison analysis between all electric vessel and traditional diesel engine.

The performance of achievement by the lithium-ion battery technology in the last 5 years is tremendous. There is a great level of advancement in the level of safety, the design life, the knowledge of how to manage the battery systems, the instrumentation of the Battery management Systems for the monitoring the temperatures, heat dissipation and State of Charge. The current state of the art in lithium-ion battery technology can now compete financially with diesel-generated electricity by utilizing even lower cost shore power. The payback is direct and without any subsidies or social costs factored in.

In the coming next 10 years with the level of advancement in technology and Li-ion Batteries will lead to high energy density battery capacities, low costs and exponential grown of all electric vessels.

7. ACKNOWLEDGEMENT

1. Mikael Odland, Norwegian Electric Systems, 4th Jan 2019

2. Recayi Pecen & Michael E. Hay, 'Design and Implementation of Solar Electric Boats for Cleaner U.S. Waters', ASEE, 2005.

8. AUTHORS BIOGRAPHY

Upendra Malla holds the current position of Naval Architect at Hornblower Cruises and Events. He is responsible for the design, engineering and construction support for the Hornblower fleet. did his Master's degree from University of Michigan-Ann Arbor (2007 -2008) and undergraduate Naval Architecture from Andhra University College of Engineering (2003 -2007). He has 11 years of experience in Engineering and Management of various projects in the marine and offshore industry. He worked for International Register of Shipping, SBM Offshore Group, KBR-Granherne and Highsea Naval Architects

THE LIFE PLATFORM: LEVERAGING A VIRTUOUS CIRCLE DESIGN

M. Magherini; A. Meredith Hardy and **A. Thoumazeau;** Lateral Naval Architects, UK

SUMMARY

The end of 2018 saw the delivery of the motor yacht Bravo Eugenia. She is built by Oceanco based on the LIFE platform developed in collaboration with Lateral Naval Architects. This study defines the LIFE platform (i.e. Long, Innovative, Fuel efficient, Eco conscious) its key features and outlines its fundamental design pillars. The relative merits of main parameters are assessed against a typical similar sized yacht by use of baseline fleet data. LIFE's low weight in relation to length and reduced powering requirements combined with a hybrid propulsion system result in a virtuous circle at which the centre is the single tier engine room. The benefits of the single tier layout such as the higher flexibility in designing the accommodation spaces and the improved balance between technical and luxury areas are outlined. In this context, various designers' LIFE platform interpretations are provided as examples for discussion. The LIFE hybrid propulsion system is presented and the various modes of operations for which the hybrid system is conceived are discussed.

NOMENCLATURE AND ABBREVIATIONS

Δ	Displacement [t]
B/ BOA	Beam Overall [m]
CE	Admiralty Coefficient
CPP	Controllable Pitch Propellers
C_R	Residuary Resistance Coefficient
$DISPV_{1.025}$	Volumetric Displacement in seawater
EPM	Electric Propulsion Motor
ER	Engine Room
F_N	Froude Number
GM	Metacentric Height [m]
GT	Gross Tonnage
Hs	Significant Wave Height [m]
L/ LOA	Length Overall [m]
LWL	Length Waterline [m]
ME	Main Engines
NT	Net Tonnage
P_S	Shaft Power [kW]
RMS	Root Mean Square
TBO	Time Between Overhauls
Tp	Roll Period

1. INTRODUCTION

Traditional motor yacht design often begins with the conception of interior layouts, or an exterior profile. Following the development of these, the naval architect is left to determine the resulting gross tonnage and optimise a hull and propulsion package around the intended design.

Sailing yacht design requires a different approach. Given that the available power is limited by the size of the sail plan, the heeling moment and the natural forces available, a holistic approach is required. Whilst length may also be fixed, the sailing yacht naval architect needs to pay far more attention in reducing the weight in relation to length, so naturally the GT of sailing yachts is far lower than that of a comparable sized motor yacht. A design that performs well will need to have a balanced approach between volume and weight compatible with performance of the chosen length.

This is in contrast with motor yacht design where the designer's role is usually separated from the naval architect, with the former exerting greater influence over the GT. The naval architect is left to optimise where possible, but ultimately if more speed is desired, it is generally a case of selecting bigger engines.

The LIFE platform (i.e. Long, Innovative, Fuel efficient, Eco conscious) delivers a design that operates in a virtuous circle: less volume, less weight, less power, less technical space, in turn reinforcing the need for a lower overall gross tonnage and afford a better balance of technical and luxury space on-board.

In the frame of this analysis, it is worth highlighting that the LIFE design platform is not a pre-engineered hull of fixed dimensions and characteristics. LIFE is rather a technical platform; a defined solution space within a feasible range of dimensions that leverages a naval architecture philosophy and a group of technologies upon which a fully custom design can be developed.

2. LIFE PLATFORM DNA

Inspired by the more holistic and performance orientated approach that is evident in sailing yacht design methodology, LIFE platform is conceived upon 2 key naval architecture fundamentals:

- Weight in relation to length – for any given weight increasing length is favourable (high length-displacement ratio)
- Speed in relation to length – for any given speed increasing length is favourable (lower F_N).

Some modern motor yachts operate at high $LWL/\Delta^{(1/3)}$ ratio (extended length, aluminium structure combined with lightweight interiors). However, the critical difference is that typically other designs have used the high-level hydrodynamic performance to increase top speed, whereas the LIFE platform uses it to reduce engine size. It is worth highlighting that LIFE platform DNA is based on a steel hull, aluminium superstructure and no use of particular lightweight materials. I.e. it is the architecture of the platform, not the use of expensive materials and outfit that bring the benefits discussed.

Recent studies on yacht operational profiles [1, 2] have concurred that motor yachts statistically spend only 2% of their total sailing time at top speed. This seems to unveil the likelihood of selected top speeds of most displacement motor yachts to be a vanity choice rather than being based upon true operational demands.

Following this approach, the LIFE platform top speed is kept within the lower range of similar GT yachts whilst increasing the L (i.e. decreasing F_N rather than keeping it constant by increasing speed). Whereas typical large displacement yachts will be operating between a $F_N = 0.34 – 0.36$, LIFE's selected top speed gives a $F_N \leq 0.32$. This results in the yacht operating in a less steep area of the resistance curve avoiding excessive additional power requirement for a small increase in top speed. This contributes greatly in reducing overall power.

The initial LIFE platform range of dimensions are shown below. The LOA is for a plumb/vertical style bow. Raked bows will see higher LOA possible:

- LOA = 102 - 112m
- Gross Tonnage = 2950 – 3400
- Top Speed = 19 knots

Figure 1 (at the end of this paper) allows for a comparison of the LIFE platform LOA and GT against existing motor yacht vessels. For this analysis, reference is made to fleet statistics about LOA and GT known from previous projects and publicly available data.

Focussing on the top end of the LIFE platform size and to demonstrate the interesting proportions of the design, it can be seen that:

- a typical 112m yacht has approximately a GT of 4900; this is 1500 more than the LIFE platform.
- a typical 3400 GT yacht has a LOA of approximately 95m. This is 17m shorter than the LIFE platform.

LIFE is designed with lower GT/LOA ratio compared against typical fleet vessels. Typical $LWL/\Delta^{(1/3)}$ values for displacement motor yachts in the range of 80-120m are shown in Figure 2: LIFE platform operates at higher ratios, resulting in a significant decrease in resistance.

Having stretched a typical design in length, whilst maintaining GT, the result is a reduced profile area but also a decreased BOA, which in turn leads to further lower resistance driving down the size of required propulsion machinery. This further improves weight in relation to length.

This is the tipping point: the design effectively enters in a virtuous circle of reducing power and weight.

From this analysis, it is evident that a basis for comparison needs to be addressed.

3. BASIS OF COMPARISON: GT

In the evaluation of commercial vessels, it is common to refer to the GT value (or NT) as one of the key design parameters for the prospective owner / operator in order to demonstrate size and cargo capacity. Similarly, it can be argued that in the evaluation of a yacht, in addition to the subjective aesthetic considerations, the GT should be considered equally as a defining factor for the owner ensuring a set level of luxury area for enjoyment. The paradigm shift for assessing the relative merits of the LIFE platform against other vessels is to conduct a comparative analysis against similar GT vessels.

4. BENEFITS OF MAIN PARTICULARS

In the following sections, this paper compares the relative merits of the LIFE platform main particulars against a typical vessel at identical GT of 3400. In this context, the baseline motor yacht main particulars are determined by statistical data of existing large yachts. A trendline is used to represent yachts of typical proportions and form. As previously shown, the LOA would be 95m.

The baseline 95m vessel is initially developed from an existing selected motor yacht where the main particulars and performance are known to the Authors, and close to the target values for comparison. Next, the vessel is scaled using the following assumption:

- L, B, and T are slightly increased to match target GT according to trend fleet data [3]
- Half load displacement has been scaled by the difference in GT when considering the powering
- Internal and external deck areas were scaled according to both L*B ratio and GT ratio of the selected candidate vessel. The more conservative values were chosen for comparison.

In the frame of the comparative analysis (Sections 4.1 (a), 4.3, and in Chapter 6) the LIFE design point is set at LOA=110m with GT=3400. This design point was selected as mature data was available within the Lateral office for a design of this size for comparison purposes.

4.1 RESISTANCE AND POWERING

The application of the Admiralty Coefficient (CE) against Speed Coefficient is regarded as a valid benchmarking system in assessing hydrodynamic performance of hull forms.

Admiralty Coefficient and Speed Coefficient are respectively defined as:

$$CE = \frac{0.7477 * V^3[knots](DISV_{1.025})^{\frac{2}{3}}[m^3]}{PE[kW]};$$

$$Speed\ Coefficient = \frac{V\ [knots]}{DISPV_{1.025}^{\frac{1}{6}}}.$$

CE seems to compare lines plans on a fairer basis rather than strictly looking at other coefficients (such as C_R): vessels with high CE achieve high hydrodynamic performance (i.e. less power required to reach a given speed * displacement product).

Figure 3 shows comparison of LIFE platform CE against other displacement hull forms of more typical main dimensions achieving various level of optimisation.

Based on the CE comparison, it is evident that LIFE's main particulars result in a high level of hydrodynamic performance.

4.1 (a) Fuel Consumption

LIFE's low resistance reduces the amount of fuel required to achieve a target range against fleet typical values.

Figure 4 reports the fuel consumption of the LIFE platform against a typical 95m motor yacht: 30% fuel reduction is achievable across the speed range.

It is worth highlighting that comparison is made for diesel mechanical propulsion trains: LIFE is conceived with a hybrid propulsion system (described in Chapter 5), hence further fuel saving in the low speed range (i.e. non-optimal engine rpm/MCR) are also achievable.

For the same range, the resultant reduced tankage and fuel oil deadweight further contributes to the virtuous circle with reduced displacement.

4.2 SEAKEEPING

The inherent DNA of the LIFE platform and resultant main dimensions, low profile shape and VCG can have a valuable effect on the seakeeping behaviours of the yacht.

Underway, the long LWL and deep forefoot provide good seakeeping performance in terms of reducing pitch motions and accelerations in large head seas.

The low beam provides opportunities for a low GM value compared to typical fleet values. During design a GM in the region of 1.3 – 1.6m was targeted. Low GM results in longer natural roll periods and tender motions (reduced roll acceleration). When coupled with an effective stabilisation system to reduce roll amplitude a high level of comfort can be achieved. However, this exercise of reduced GM is limited by compliance with regulatory stability requirements.

No single parameter can be used to define seakeeping performance of a specific design. However, it is common practice on yachts to benchmark the stabilised roll performance with absolute roll angle at zero speed in resonant beam waves against existing fleet data.

In setting a comparative analysis, a candidate reference motor yacht was selected based on similar natural roll periods and low GM. Experimental tests with scaled models on both LIFE platform and the reference yacht were conducted at MARIN (Maritime Research Institute Netherlands); hence, a high level comparative analysis can be undertaken.

Vessels	**Tp [sec]**	**GM [m]**	**Fins**	**Area [m²]**
LIFE	9.2	1.56	2x	21.7
Ref. yacht	9.5	1.36	4x	26.0

Table 1: [LIFE main seakeeping particulars against selected reference vessel].

Table 2 reports RMS for roll angle at different Hs in the beam sea zero speed condition (irregular waves). Both models are fitted with bilge keels, and have identical stabiliser suppliers and systems.

Hs	**LIFE**	**Reference vessel**
[m]	**RMS Roll [deg]**	**RMS Roll [deg]**
0.5	0.29	0.25
1	0.71	0.65
1.5	1.58	1.40

Table 2: [RMS Roll comparison with active fins].

It can be seen that despite the smaller fins, and also the higher test GM, the LIFE platform achieves results close to the reference yacht.

The low block coefficient of the hull form and round bilges does have the advantage that 2x stabilisers can be positioned close to midships (where they are most effective) rather than 4x stabilisers positioned fore and aft. The smaller fins also reduce resistance and power, further enhancing the virtuous circle philosophy.

4.3 EXTERIOR DECK AREAS

Scaling the deck areas following the assumptions stated at the beginning of this chapter, allows for direct comparison at constant GT.

Parameters	LIFE	Baseline 95m
GT	3400	3400
LOA [m]	110.0	95.0
External deck area [m^2]	1008	853

Table 3: [External deck area comparison between LIFE against typical 95m MY].

The baseline 95m motor yacht external deck area is 155m^2 less (approximately 16% decrease) than the LIFE platform's areas. This can be beneficial in cases where space for outside living, or compliant helipads is of particular interest to owners.

5. HYBRID SYSTEM

On top of an efficient hull form and reduced powering requirements, the LIFE platform benefits from a hybrid propulsion system powering a twin shaft train fitted with controllable pitch propellers (CPP).

A hybrid system by definition uses a combination of different energy sources or propulsion devices.

LIFE presents a diesel electric hybrid system also including batteries, which is designed around a direct drive main engine, coupled to an electric motor by dual input gearbox (see Figure 5). The electric motor can operate as both a motor and a shaft generator.

Typical advantages of hybrid system against a typical diesel mechanical system are:

- Operational flexibility for control of the yacht;
- Potential to operate with the lowest fuel use and carbon emissions;
- Reduced total installed power of prime movers;
- Maximised TBO (Time Between Overhaul) by maintaining the best loading on equipment;
- Flexible arrangement of main propulsion and power components allows for low profile machinery spaces;
- Single tier engine room is feasible (detailed in next chapter).

On the other hand, typical disadvantages compared to a traditional yacht propulsion system to consider are:

- Increased system complexity.
- More machinery volume required (at same P_S).

5.1 BATTERIES

At the time of writing, it has been demonstrated [4] that the current most practical use of batteries on board large yachts is found for peak shaving and for electrical load management which is a highly effective method of delivering reductions in generator sizing. In addition, optimum generator loading can be achieved across a wide range of operating conditions with increased security of electrical power supply.

The size and weight of batteries required to achieve this on the LIFE platform is consistent with the overall philosophy of the virtuous circle design.

As battery energy density and performance improves, increased periods of "battery only power" and "battery only propulsion" can be expected on future LIFE platform designs.

5.2 MODES OF OPERATIONS

The key benefits of employing a hybrid system consists in combining the advantages of both diesel-electric and diesel mechanical systems.

The following are typical modes of operation for diesel electric hybrid systems. All these factors need to be carefully considered alongside the yachts likely operating profile.

5.2 (a) Slow speed

The mode will run with electric propulsion (only generators & EPM operating). Speeds of around 11-12 knots are possible.

This mode allows generators to be optimally loaded. Main engines are switched off avoiding poor low loading. Very low level of noise and vibration is produced; all the exhaust emissions are routed to the mast via soot filters avoiding black marks and cleaning.

5.2 (b) Eco Cruise speed

The main engines are running with the electric motors operating as shaft generators (i.e as a PTO system). All the generators are switched off to reduce wear and maintenance and increase their TBO.

As the main engines are supplying both the propulsion and hotel power, the load is sufficient that they can be made to work at their optimum efficiency. This can be ensured by adjusting the propeller combinator curve to suit engine performance.

5.3 (c) Fast Cruise speed

At high cruise speeds, the system can operate in a traditional diesel-mechanical mode (ME only for propulsion, generators for hotel load).

5.3 (d) Boost speed

For boost speed, all the installed power of the main engines and generators is being used to deliver the top speed. The system will operate as a diesel electric and diesel mechanical system combined (a boost is provided to the shafts from the generators via the EPM).

6. SINGLE TIER ENGINE ROOM

The virtuous circle (high $L/\Delta^{(1/3)}$ ratio, moderate powering requirements, hybrid propulsion) results in the ability to select a lower main engine block series, which in turn means a significant reduction in the vertical height required in main machinery spaces.

Following this philosophy unveils the centre of the virtuous circle: a single tier engine room. Where conventional designs of similar GT would employ a twin-deck engine room, LIFE platform is able to utilise a single tier design.

For instance, typical motor yachts of this size could employ MTU 4000 series or equivalent engines, whereas LIFE only needs to utilise the 2000 series which is significantly smaller and lighter.

Main engine dimensions (length*width*height) for both LIFE and the reference vessel are reported in Table 4: ME length and height are significantly reduced compared to higher engine block series.

Parameters	LIFE	Baseline 95m
GT	3400	3400
LOA [m]	110.0	95.0
ME [mm]	2260*1320*1460	4040*1470*2440
Luxury area [m^2]	903	805

Table 4: [Comparison LIFE against baseline 95m MY].

Moreover, Table 4 shows internal luxury areas for the baseline motor yacht compared to the life platform. The single tier engine room results in a more optimal use of luxury space giving approximately an extra 100 square meters compared to a twin tier engine room vessel of same GT.

Figure 6 shows an artistic impression of the hybrid system and the single tier ER. The highlighted volume indicates the area gained for accommodation spaces at Lower Deck level. At such location, it is unusual for typical yachts of this size to have luxury space: this creates design opportunities that are covered in Chapter 7.

7. LIFE PLATFORM DESIGN ASPECTS

The virtuous circle established in the LIFE platform DNA results in some key recognisable design features:

- Stretched length gives sporty/ slender/ light yacht look rather than a traditional voluminous superyacht;
- Single tier engine room gives higher flexibility of internal spaces, as such:
  - Lower deck offers more accommodation spaces, (guests or crew) against similar GT vessels;
  - Circulation between aft areas (e.g. beach club) and mid-ship is improved (e.g. main stairwell / access to crew areas).

The following sections report a variety of LIFE platform interpretations by various yacht designers.

7.1 BRAVO EUGENIA: NUVOLARI LENARD INTERPRETATION

Bravo Eugenia, is the first delivered vessel conceived from the LIFE platform [see Figure 7]. She is at the higher design size range of platform and was recently presented with the Technology and Innovation award at the 2019 Yacht Club Monaco's La Belle Class Explorer award [5].

On Bravo Eugenia, six guest cabins are located at lower deck level above the single tier engine room. As such, guests are next to the aft beach lounge/ gym, SPA areas and water toys garage. Circulation and guest experience around the vessel are enhanced.

7.2 BALANCE: SINOT INTERPRETATION

Balance [Figure 8] was presented during Monaco Yacht Show 2018.

The 4 guest cabins above the single tier engine room at Lower Deck gives easy access to the Laguna lounge/ cinema theatre, the gym /SPA and to the water toys in the aft areas of the vessel.

Two VIP cabins are situated at the Main Deck with the deck above being reserved to Owner.

7.3 VIRTUS: MC PHERSON INTERPRETATION

With L = 102m, and 5 decks, Virtus motor yacht [Figure 9] is at the lower range of the LIFE Platform dimensions both in terms of length overall and GT ($\leq$3000).

All six guest suits are located at lower deck above the single tier engine room, granting proximity to the aft beach club/ cinema area. Owner's area is at Bridge Deck within this configuration.

8. CONCLUSIONS

This paper analyses the merits of a motor yacht platform designed by following a holistic approach.

As is often the case, it is Authors' opinion that major innovations succeed by leveraging existing technologies in a novel combination.

It is arguable that, in recent years, other superyachts have been designed and built by targeting low volume, and a light weight against length approach; however, the enhanced level of hydrodynamic performance has been typically exploited to fulfil the constant seek of higher top speeds.

On the contrary, the LIFE platform's key achievement consists in the unique combination of leveraging first naval architecture principles whilst taking a pragmatic approach to operational profiles. As such, this paper demonstrates how the LIFE platform can achieve a significantly lower level of fuel consumption compared to a typical superyacht fleet vessel whilst also offering more luxury/ accommodation space as well as external deck space. This is done without the need for aluminium hull construction of light weight interiors.

LIFE's virtuous circle has been analysed and comparisons made against a baseline motor yacht are shown. Following this, with the aim of summarising the key features of a LIFE platform design, the following pillars are identified:

- Optimised main particulars to increase length-displacement ratio whilst keeping a moderate Froude Number at top speed;
- Hybrid propulsion system;
- Single tier engine room.

This study shows that these key features respectively result in:

- Higher hydrodynamic performance (i.e. less power required to reach a given speed * displacement product);
- Operation flexibility and potential for low fuel use and carbon emissions with maximised machinery TBO;
- Optimised use of luxury space compared to a typical double tier engine room by creation of additional accommodation space in an unusual location.

The inherent DNA and approach of the platform results in outcomes which are synonymous with an Eco Conscious design. Although not the topic of this paper, it is worth to note that the platform is also configured with large waste heat recovery systems on the generators, IMO Emissions Tier III compliance (prior to the implementation date requirement) as well as other green technologies.

The current study does not address practical aspects of adopting hybrid propulsion systems (i.e. system commissioning and integration, required higher crew technical skills to operate the system correctly, etc.) This will be motivation for future work on the subject.

9. REFERENCES

1. VAN LOON, P.T.; VAN ZON, P.; "From Operational Profile to Hybrid Propulsion", *HISWA Symposium*, 2016.
2. BREALEY, S.; ROY, J. and HOPPE, F.; "Advanced Propulsion Trains for Superyachts", *HISWA Symposium*, 2014.
3. LATERAL NAVAL ARCHITECTS, extract from "*Large Yacht Statistics Issue VII*", 2018.
4. BREALEY, S.; "The Future of Battery Energy Storage in Large Yacht System", *HISWA Symposium*, 2018.
5. https://www.oceancoyacht.com/en/press/press-releases [accessed on 25/03/2019].

10. AUTHORS BIOGRAPHY

Matteo Magherini (CEng, MRINA) holds the position of Naval Architect/ Business Developer Coordinator at Lateral Naval Architects. He currently works in a number of yacht projects from initial development to engineering phase. He is also involved in the provision of wider company consultancy services and business strategies.

Alex Meredith Hardy (MEng, MRINA) holds the position of Principal Naval Architect at Lateral Naval Architects. In this role he looks after the naval architect team and capabilities of the company. He was involved in the development of the LIFE platform since the inception of the project.

Adrien Thoumazeau holds the position of Senior Naval Architect/ Research and Development Coordinator at Lateral Naval Architects. He is involved in wide range of yacht projects from concept design stages through to the detail design stages as well as coordinating innovation. The development of the LIFE platform was one of the first projects he worked on at this company.

11. ACKNOWLEDGMENTS

The Authors wish to extend the merits of the LIFE platform development with Oceanco, especially the Project Development Team, and the MARIN (Maritime Research Institute Netherlands) Team.

LIFE platform is an example of a collaborative approach where the result exceeds the sum of the individual parts.

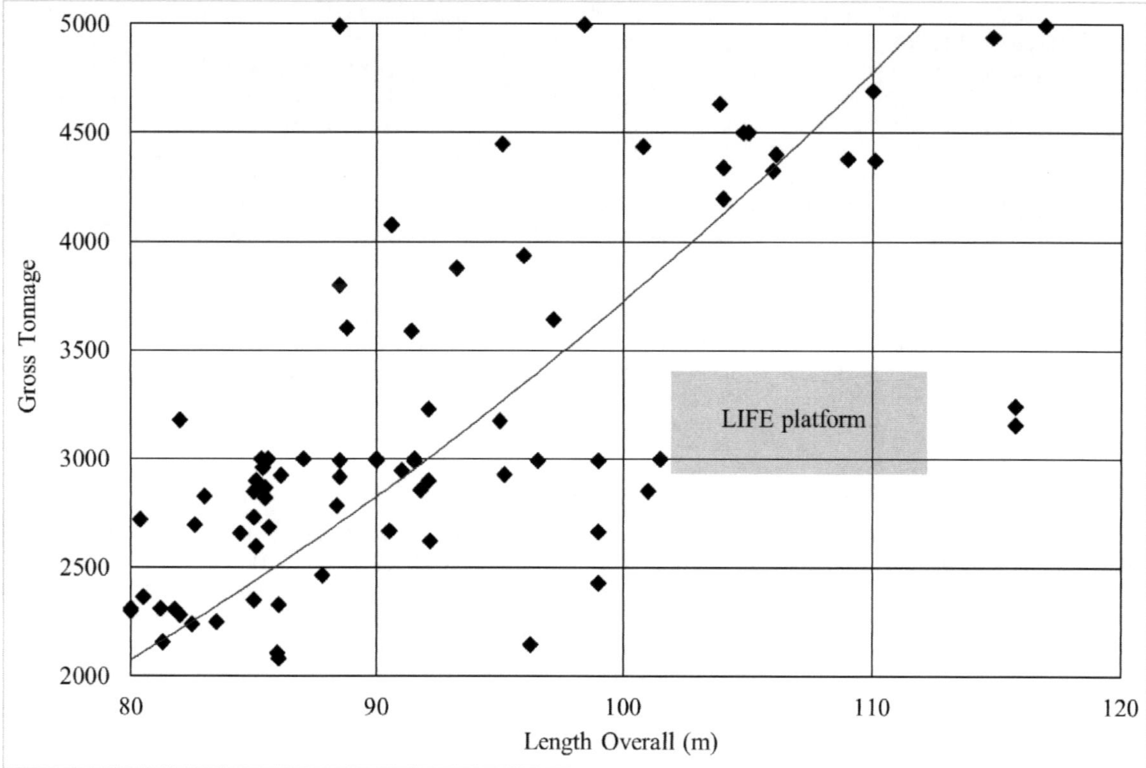

Figure 1: [GT vs LOA of existing yachts] [3].

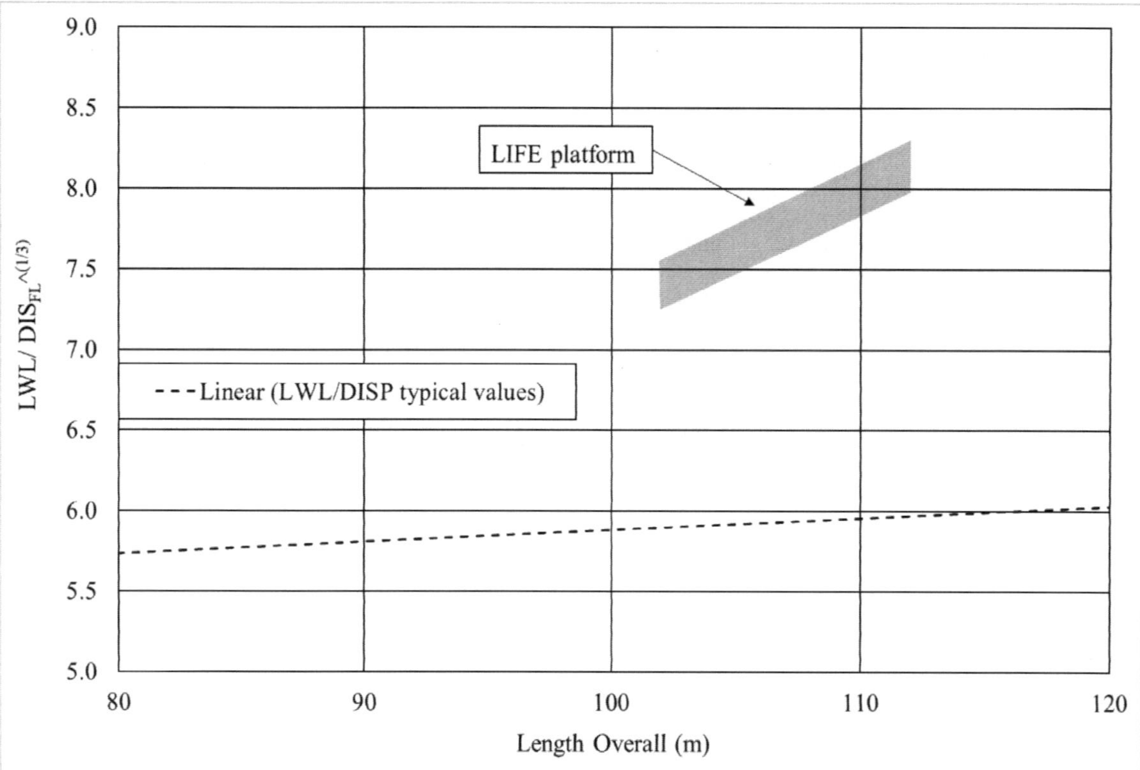

Figure 2: [LWL/$\Delta^{(1/3)}$ ratio at Full Load vs LOA compared to trend of existing yachts] [3].

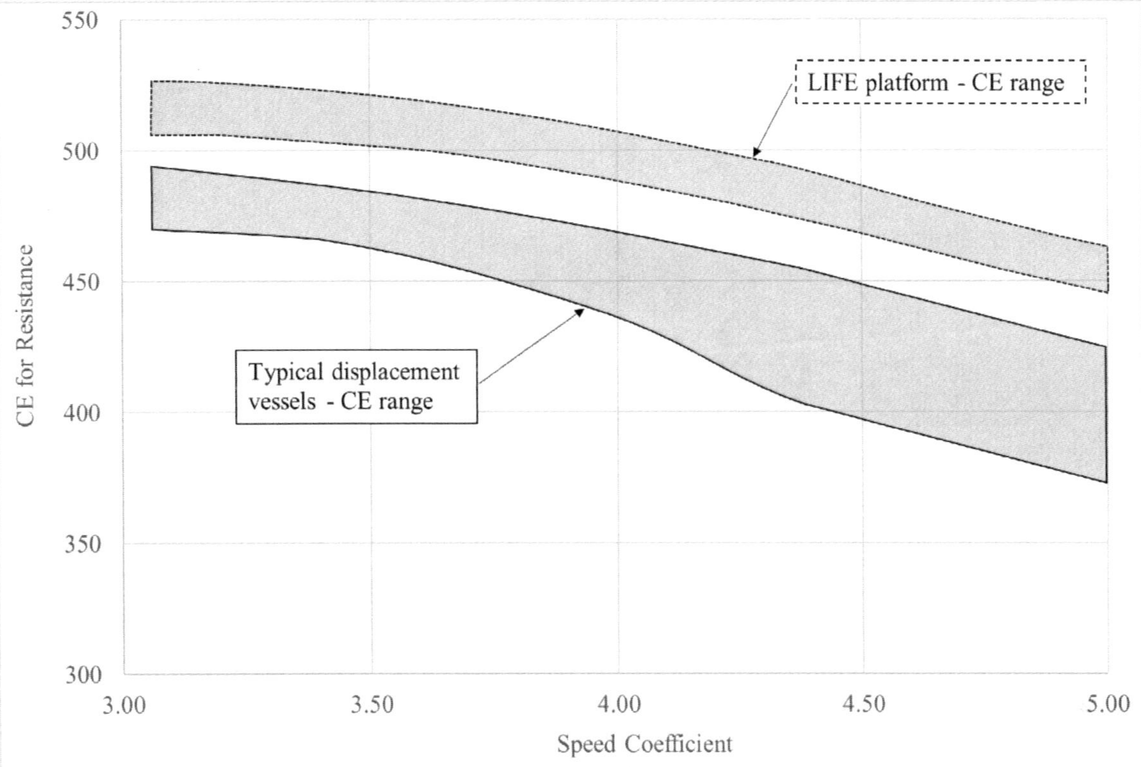

Figure 3: [LIFE platform CE comparison against other vessels].

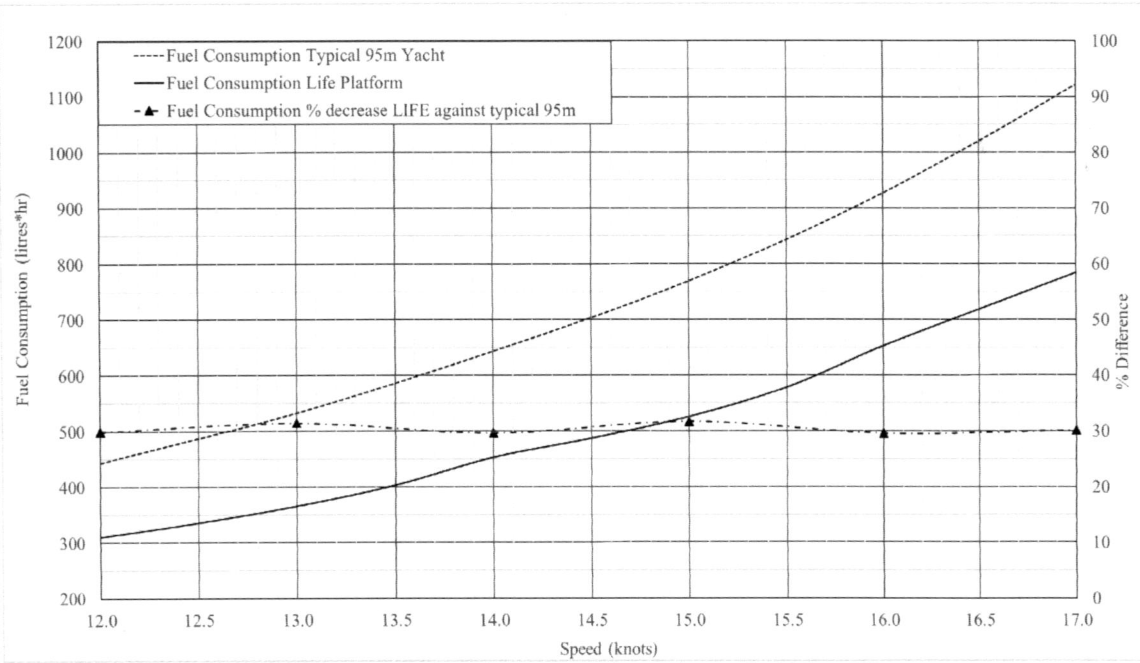

Figure 4: [LIFE platform fuel consumption against typical 95m MY].

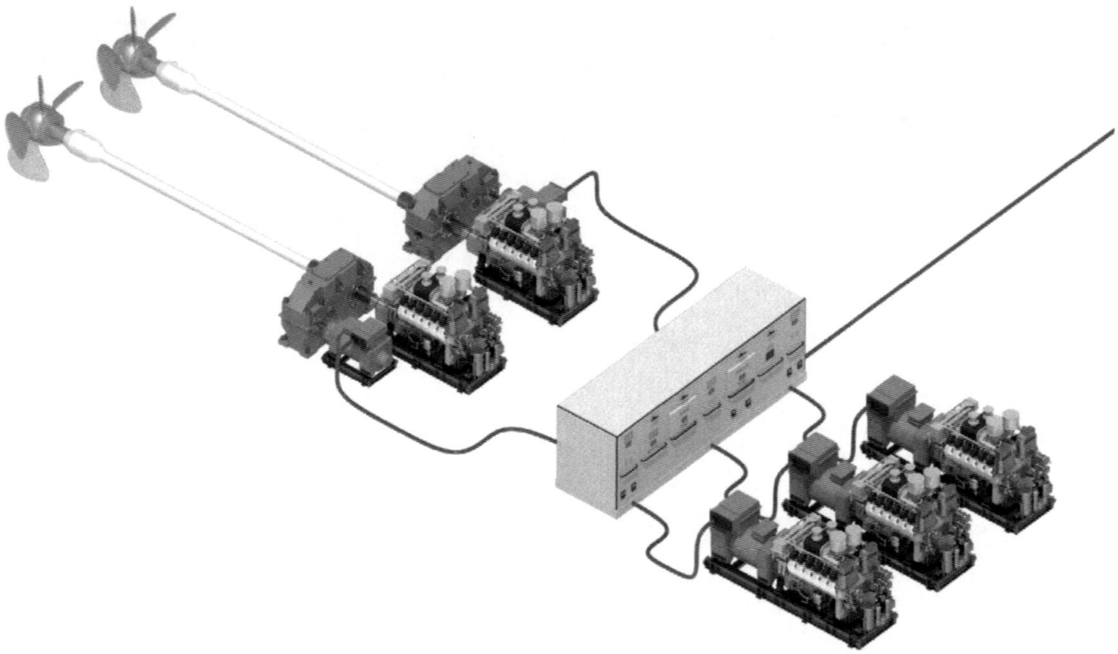

Figure 5: [Artistic schematic of typical diesel-electric hybrid system]

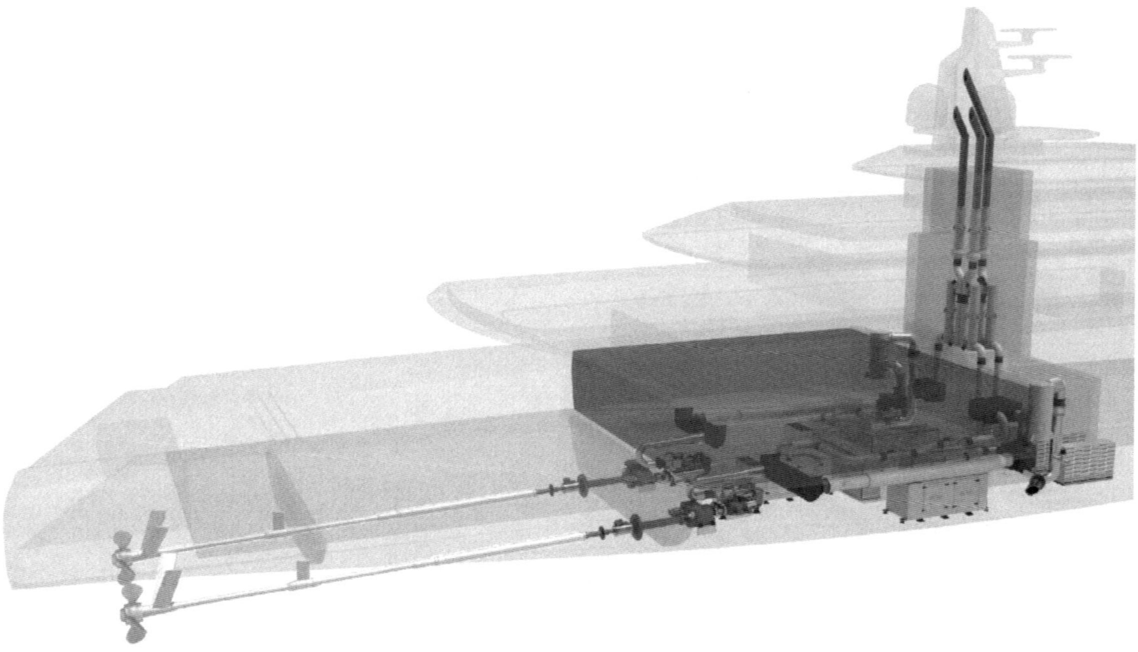

Figure 6: [Artistic impression of LIFE platform single tier engine room. Highlighted volume shows extra accommodation spaces at Lower Deck.]

Figure 7: [Bravo Eugenia (photo credit: Francisco Martinez)].

Figure 8: [102m Balance (rendering: Sinot Yacht Architecture & Design)].

Figure 9: [102m MY Virtus (rendering: Mc Pherson Yacht Design)].

PREMIUM YACHT PROPULSION FOR EXCLUSIVE PERFORMANCE

Andreas Witschel, SCHOTTEL GmbH, Germany

SUMMARY

SCHOTTEL offers a wide range of main and auxiliary propulsion systems for yachts. This paper will present characteristics of different propulsion solutions: The 5-bladed controllable pitch propeller for highest propulsive efficiency for changing speed or load or the EcoPeller with best values in overall efficiency and course stability. It provides unprecedented performance and ensures that its owners benefit from low fuel consumption and thus reduced emissions and operating costs.
Thanks to an installation flush with the hull, the pump jet is ideal for operation in extremely shallow waters. It can be used as manoeuvring aid, additional booster or take-home device. Low-vibration and low-noise rim thruster enables more comfort on board.

1. INTRODUCTION

Each vessel type has special requirements to the propulsion and manoeuvring systems. Functional market segmentation helps to target the exact needs of customers. SCHOTTEL engineers and technicians have developed the appropriate solutions also for yachts.

2. AUXILIARY PROPULSION

2.1. THE SCHOTTEL PUMP JET (SPJ)

The SCHOTTEL Pump Jet is equipped with an impeller aspirating the water axially and ejecting the water radially, similar to a dredger pump wheel. Such an impeller is suitable to increase static pressure in jet housing much more, compared to an ordinary Kaplan propeller. In combination with the three-dimensionally extending diffuser, static pressure is further increased.

See principle below:

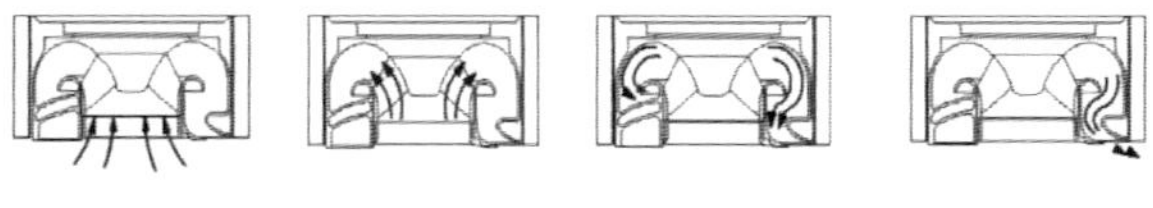

The water flow leaving the housing with a downwards angle of 15° is accelerated by the three nozzles to a speed of 60 – 70 km/h. The area of all three nozzles together is one third of the area of the large suction inlet. Therefore also water entry speed is a one third of the exit speed in the nozzles. The risk of aspiration of elements from the bottom in shallow water is the low inflow speed reduced. The thrust is 360° steerable and full thrust force is available in all directions.

The pump jet can be installed flush into the hull, so that it is fully integrated into the ships body and it does not increase vessels resistance. Additionally, there is no risk of damage due to floating debris in water.

The important difference to all other 360° steerable jet propulsion systems available on the market: The pump jet operates with low volume flow and high jet speed to produce/generate thrust.
Due to low volume flow, the pump jet is characterized by the best thrust / size ratio on the market. Thanks to its space saving installation, it can easily be integrated into vessels with limited space. Because of this it is also suitable for retrofits.

The water outlet angle is only 15° to downwards. Thus nearly the entire thrust force is effective in direction of travelling. There is no risk for swimmers.

Due to the high jet velocity, the pump jet can be used for maneuvering up to vessel speeds of 12kn, a high amount of thrust even at full speed remains available. Other jet systems equipped with Kaplan propellers and operating with high volume flow show at same input power already at speeds above 2kn less of thrust performance. This is explained by the lower water jet speed but also mainly due to much lower suction performance of Kaplan propellers compared to the SCHOTTEL Pump Jet impeller.

The pump jet is the ideal slow speed maneuvering device for vessels with hulls characterized by good advanced ratio as yacht hulls. Already by applying relatively low input power a relatively high sailing speed can be reached. That is why the pump pet is in operation on more than 100 yachts as electric driven stern thruster to achieve a hybrid propulsion concept. Electric board net power enables slow speed maneuvering and slow speed sailing, while high speed propulsion is provided by diesel direct shaft lines.

The pump jet is installed into passenger vessels, luxury yachts as well as research vessels and replaces sometimes also transverse thrusters with the advantages of additional DP capability and extremely silent operation. The low level of structural borne noises never can be met by any tunnel thruster or other jet systems.

See example below:

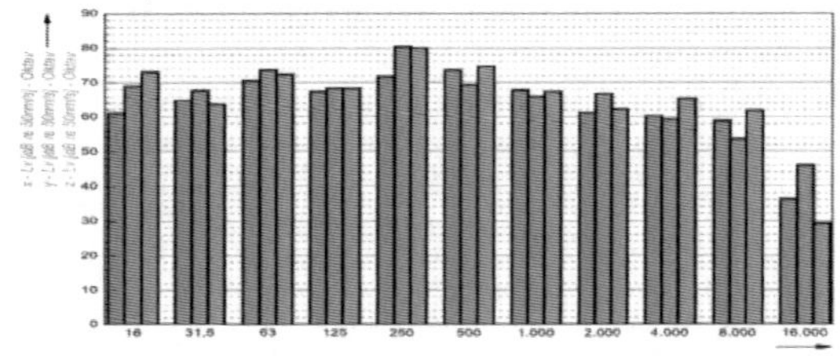

Structure borne noise measured n x,y and z axis over frequencies in Hz for SPJ 132 RD with 477 kW input power

2.2 THE SCHOTTEL RIM THRUSTER (SRT)

The product development of the SCHOTTEL Rim thruster series started in 2012 and took almost six years before the first unit was delivered.

Several prototypes were built and extensively tested regarding corrosion behaviour, regarding wear and tear of friction bearings in sediment rich waters before entering into the market.

The SCHOTTEL Rim thruster has been developed with entire focus on yacht requirements. It is equipped with a permanent magnet ring motor which is driving the RIM with propeller blades showing to centre of tunnel. The stator of the permanent motor is housed into water tight housing made of Ni-Resist and maintained on higher pressure than the outside sea water pressure to avoid water to enter. Additionally, the stator windings are casted into a water resistant and heat conducting resin. The thruster needs no lubrication oil or grease and operates with axial and radial friction bearings lubricated by sea water. It fulfils the highest environmental restrictions. All rim thruster materials have been selected with the focus to avoid the risk of corrosion. Therefore, stainless steel is used for the entire friction bearing support system.

The rim thruster is supported by elastic brackets inside the tunnel. These brackets serve as well for feed through power cables, control cables and compressed air connection.

The friction bearings can be exchanged on the vessel. The stainless steel bearing supports are segmented and dismountable.

The rim thruster is extremely silent in operation as the ring propeller generates no pressure pulses. A very expensive sound damping measures on yacht builder's side can be saved when a SCHOTTEL Rim thruster will be installed.

The focus in design of the SCHOTTEL Rim thruster series was laid on shortest possible thruster length to allow installation most forward in V-shaped bow and narrow skegs. Hence a 200 kW SRT 800 can be installed into a skeg of only 500 mm width.

Here below an over view table of the advantages of SCHOTTEL Rim Thrusters in comparison to conventional tunnel thrusters:

Noise and Vibration

Franz-Achim Arntzen

SRT	Tunnel Thruster
(+) ring motor in wet area (+) elastic mounted in vessel structure (+) low cavitation by gapless blade design (+) homogenous flow to propeller due to missing gearbox (+) less investment on shipyard's end for sound damping measures inside ship	(-) thruster motor inside the vessel, additional source of noises (-) noise/vibration induced into vessel structure due to pressure pulses of propeller blades (-) high cavitation risk between propeller blade tips and tunnel (-) by gearbox disturbed flow to propeller. Skew reduces this effect, but doesn't eliminate it

Dimensions and Weight

SRT	Tunnel Thruster
(+) compact ring PM motor (+) narrow SRT width allows minimum tunnel length (skeg installation) (+) compact design allows longer lever to ships turning point, because SRT can be mounted more forward with shorter tunnel (+) more space available in ship because bow thruster room needs less height or length because no separate electric motor has to be accommodated (+) less of loss of displacement due to shorter tunnel in bow	(-) additional motor inside the vessel. Additional space needed for motor handling and servicing (-) much higher system weight (-) longer minimum tunnel length due to gear box (-) higher tunnel length requirements are forcing the tunnel position towards the vessels center (-) tunnel in tunnel design (super silent) requires to locate extremely high tunnel diameters in fore ship compared to installed power

Environment-friendly design

SRT	Tunnel Thruster
(+) energy efficient system (+) sea water lubricated system, no lubricants (+) sea water cooled, no added cooling (+) low underwater noise emission	(-) danger of oil leakages at dynamic seals (-) high vibration and (underwater) noise emission

Maintenance

SRT	Tunnel Thruster
(+) no dynamics seals (+) no need for oil change (+) friction bearings are radially segmented and axially consisting of single round pads (+) friction bearing radial segments and axial pads can be exchanged without dismounting the SRT (+) friction bearing parts are even under water exchangeable (+) easy and fast de/installation at yard	(-) sealing rings are wear and tear parts and have to be exchanged from time to time (-) sealing ring exchange requires dry docking (-) oil change intervals to be considered (-) a lot of rotating parts as bevel gears and bearings increase the probability of service interventions

Hydrodynamic performance

SRT	Tunnel Thruster
(+) equal thrust in both directions (+) unique propeller blade design for significantly reduced cavitation risk (+) no gap between blade tips and tunnel increases efficiency (+) no reduction of tunnel section by gearbox. Active flow area of tunnel is increased	(-) high cavitation risk between propeller blade tips and tunnel (-) efficiency losses by flow disturbance due to gearbox in tunnel

2.3 SWING OUT AND RETRACTABLE RIM THRUSTER WITH AZIMUTH FUNCTION (SRT R)

Currently SCHOTTEL develops for the power range of 315 kW a swing out and azimuting rim thruster with continuous power of 315 kW to be located into the aft-ship of yachts on CL. The product will be delivered along with a weld-in container of smallest possible dimensions. The azimuting RIM thruster is able to generate thrust into all directions (360°). However, this is done by ± 105° steering angle and inversion of motor rotation. Due to this innovative steering method there is no slip ring assembly for electric power transfer necessary, which is a very expensive and service intensive component.

3. MAIN PROPULSION

3.1 THE SCHOTTEL ECOPELLER (SRE) ®

The SCHOTTEL EcoPeller was one of the most significant developments in SCHOTTEL history.

Within three years the entire EcoPeller series has been developed in the power range between 570 kW – 3500 kW.

This was only possible due to high investments in powerful computing tools enabling SCHOTTEL to do virtual prototyping via multi body simulation with deformable bodies due to finite elements method and generating physical forces on virtual prototype by fluid dynamics calculation.

Hence all kind of tests were realized with a virtual prototype and consequently several design reviews have been made before building the first real thruster.

The reason for this development was that market showed an increased demand for most efficient thrusters for offshore and ferry applications. Besides that, SCHOTTEL remarks an increased interest for such thruster type to be installed into mega yachts with dimensions from 80 m to 120 m length.

Especially for this Mega Yacht segment the SRE 430 L FP has been developed outperforming the EcoPeller series regarding degree of slimness of lower gearbox and an increased distance of the propeller to the gear housing to maintain interaction with gear housing at lowest possible level. This allows increasing power density on the propeller of the SRE 430 L FP, allows to install reduced propeller diameters with very high propeller load and allows ship speeds over 20kn.

3.2 5-BLADED CONTROLLABLE PITCH PROPELLERS (SCP)

SCHOTTEL recently developed a new generation of controllable pitch shaft lines with the focus on military and yacht applications requiring higher power density, low pressure pulses and low under water noise.

Furthermore, a new CP hub form has been developed suitable for high ship speeds and lowest possible hub vortex and cavitation.

SCHOTTEL invested massively for CFD propeller design, MPUF3A and PROPCAV cavitation prediction tools to optimize the propeller blade design.

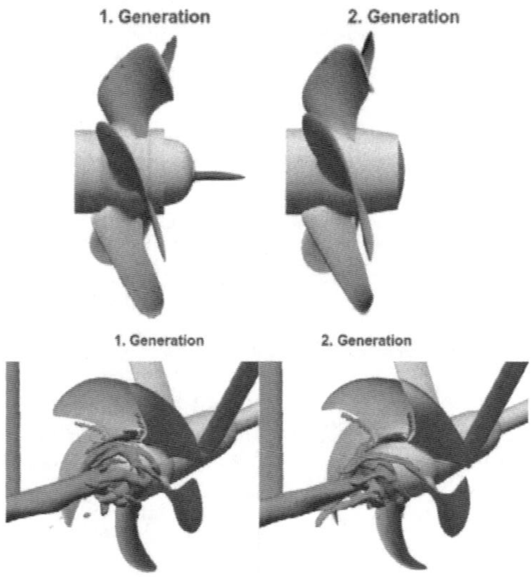

4. RESUME

With more than 30 years of crucial expertise in yacht propulsion SCHOTTEL is the ideal partner for all of your propulsion needs: The broad range of propulsion solutions – for main as well as auxiliary drives – is best suited to provide highest propulsion efficiency and high manoeuvrability while further reducing low noise and vibration levels. Thanks to its worldwide sales and service network SCHOTTEL is the ideal partner when it comes to equip yachts.

5. AUTHORS BIOGRAPHY

Andreas Witschel is a dedicated sales expert who is characterized by his extensive experience and enormous knowledge of the propulsion market. In recent years, he has held several management positions within the SCHOTTEL sales organization. The SCHOTTEL Group, with its headquarter in Spay/Germany, productions sites in Germany and China, a worldwide network of own service and sales subsidiaries is one of the world's leading manufacturers of propulsion and steering systems for ships and offshore applications. Andreas Witschel was heading the sales department for thrusters up to 1,000hp, the SCHOTTEL Application Centre and later on the Sales Competence Centre. Most recently, he was appointed Sales Director of the newly installed segment "Cruise, Ferry & Yachts".

After having graduated in Mechanical Engineering with a focus on turbo machinery and prior to joining SCHOTTEL in 2000, Andreas Witschel held various sales manager positions in internationally operating companies in machinery and plant engineering.

EXPERIMENTAL ANALYSIS TO CONTROL WELDING DEFORMATION IN THIN PLATE

M. T. Ali, Damen Shipyards Galati, Romania.

SUMMARY

Shipbuilding industries facing the problem of welding deformation for seven decades and trying to control this problem but the thin plates bring non-linear problems in the ship structure. Especially in the bulkhead of the ship structure that leads to welding deformation due to small members of the profiles. Mostly, the problem occurs in the construction of yachts building because clients need good outlook of the ship. Therefore, the development of a new effective method to reduce the welding deformation in the thin plate by optimization of welding sequences and plating method at the time welding process and this method is help to reduce the deformation and also help to increase the productivity of the shipbuilding process.

1. INTRODUCTION

1.1 BACKGROUND

Welding is a process to join metals with different arrangements of the plates. Welding induced the distortion in the thin plate. There are six types of welding distortion:

The objective of this research to analyse the welding deformation, welding deformation is caused by the welding distortion; the main factor of welding distortion is based on the following parameters which are given below:

- Longitudinal Shrinkage
- Transverse Shrinkage
- Angular Distortion
- Bowing and dishing
- Buckling
- Twisting

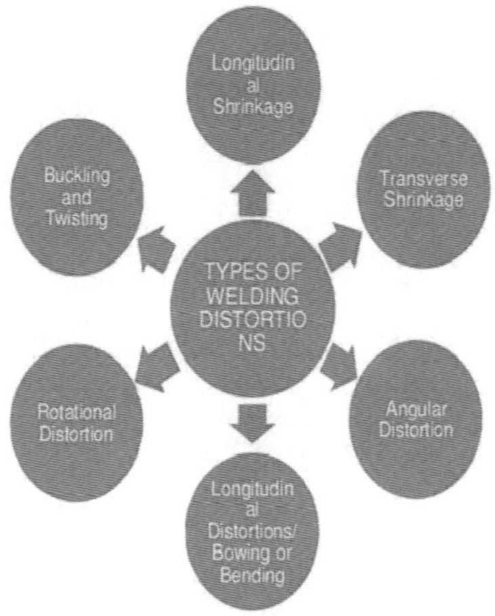

7

Figure 1: Types of Distortion in welding [2]

After welding the shrinkage is towards the weld axis longitudinally is known as ***"Longitudinal Shrinkage".***

After welding the shrinkage is towards the weld axis longitudinally is known as ***"Transverse Shrinkage".***

"Angular distortion" is the change in the shape of the plate by the welding. Consider the but welding in which the top part of the welding is wider than the bottom part.

As we know that two or more than two distortions will occur at the same time when longitudinal Shrinkage occurs then there will be the bending of the plate. That`s why is known as ***"Bowing".*** This type of Distortion happens when the weld line and Neutral Axis of the structure do not coincide.

Mostly ***"buckling"*** is occurring in the thin plate. This type of distortion occurs away from the welding area this is due to the residual stress.

"Twisting" has also happened in the thin plate this is due to the low torsional resistance. Twisting of the plate occurs when the welding is done along the centre of the member. Due to this reason, the area tends to shrink and become shorter.

1.2 WELDING

In the shipping industry, different types of arc welding are used, which are shown*:*

1. Shielded Metal Arc Welding (SMAW)
   Electrode Type
   Arc between;
   - Base metal and
   - Covered electrode
2. Gas Tungsten Arc Welding (GTAW)
   Electrode Type
   Arc between;
   - Base metal and
   - Tungsten electrode
3. Gas Metal Arc Welding (GMAW) / MIG (Metal Inert Gas Welding)
   Electrode Type
   Arc between;
   - Base metal and

- Consumable electrode

4. Flux – Cored Arc Welding (FCAW)
   Electrode Type
   Arc between;
   - Base metal and
   - Consumable electrode

5. Submerged Arc Welding (SAW)
   Electrode Type
   Arc between;
   - Base metal and
   - Consumable electrode

Cheapest and easiest welding of Grade "A" material is given below:
1. SMAW i.e. Shielded Metal Arc Welding.
2. GMAW i.e. Gas Metal Arc Welding or MIG i.e. Metal Inert Gas Welding.
3. Oxyacetylene Welding. [3]

MAG welding also is known as GMAW (Gas Metal Arc welding). It is a really good method of welding in the thin sheet but in this method, you have put extra attention at the time of welding otherwise this type of welding cause deformation and burn.

This type of welding is also known as Gas Metal Inert Gas (MIG) welding. In this welding in which shielded gas is used to heat up the electrode to join the two metal plates.

The source for this welding is constant voltage and direct current. Now, a day this type of welding process is commonly used in the industry.

1.2 WELDING DISTORTION

Welding distortion occurs due to the non-uniform distribution of the heating and cooling cycle. Due to which welding complex strain is produced during the welding. This distortion is occurred in between weld bead and base metal plate.

1.2 (a) Cause of welding distortion

The welding distortion problem in the welded plate is varied due to the grade of steel plate due to the following reasons which are given below:

1. Heat Conductivity
2. Co-efficient of the Thermal Expansion
3. Modulus of Elasticity

1.2 (b) Control of welding distortion

The welding distortion is controlled by several methods after analysing and finding ways:

1. Elastic Pretraining
2. Clamping
3. Differential Heating

1.2 (c) Checks for welding distortion

To ensure that the plate doesn`t have any welding distortion then we have to follow the following steps:

1. Before Welding Procedure and Checks
2. During Welding Procedure and Checks
3. After Welding Procedure and Checks

2. OBJECTIVE OF THE ANALYSIS

To control the welding deformation by welding sequence, a transverse plating method and clamping technique.

3. WELDING PROCEDURE SPECIFICATION

The welding procedure is the documents which contain important values of welding parameters that help the welding process. We know that computational analysis has very less control in the welding procedure; this is because of the welding procedure totally experimental procedure.

Welding procedure sequence must follow the IACS code for the welding process. Engineer and welders are responsible to follow the WPS. WPS help to produce high-quality welds.
Welding procedure specification includes the following codes:

1. API code
2. AWS code
3. ASME code
4. ISO code
5. Foreign code

Welding procedure specification has important information in the datasheet regarding material properties which are:

1. Spec and Grade
2. Yield Strength
3. Tensile Strength
4. Elongation
5. Wall thickness
6. Size i.e. diameter

WPS contain informations about the welding process which are:

1. Which kind of welding i.e. MIG, TIG or so on.
2. Method of welding i.e. Manual, Semi-automated or automated.

WPS has also the information regarding welding process parameters which are:

1. Volts, Ampere and Travel Speed
2. Polarity
3. Travel direction
4. Wire welding transfer mode
    - Spray, Short Circuit, Plasma or Globular
    - Flux Core or Shielding Gas
5. Number of welders
6. Number of passes
7. Electrode size and group number
8. Electrode AWS specification

More details are mention in the WPS which help the engineer and welder to weld the plate. [4]

By this IACS Rule # 47, we can compare the normal and high strength hull structural steel grade.

Welding procedure specification parameters for this experimental analysis are:

1. Type of welding is MIG/MAG welding.
2. The angle of the electrode is **45°.**
3. Weld bead size is **5mm.**
4. Shielding gas percentage:
    a. Argon is **82%.**
    b. CO_2 is **18%.**
5. Gas flow rate is **13 lit/min.**
6. Feed wire rate is **7.5 m/sec.**
7. Current during welding is **240 Amperes.**
8. Voltage during welding is **24 volts.**
9. Clamping time in case study # 5 is **15 – 20 min**.

4. MODELLING

The model assumes to be a temperature dependent. It means that thermal analysis is used to be a source of mechanical behaviour in the TEP-FEM analysis.

The following mechanical and thermal properties are the dependents of temperature are given below:

1. Yield Stress
2. Elastic Modulus
3. Thermal Expansion and
4. Poisson`s ratio.

For modelling of the plate need the values of temperature at the time of welding and the material of the plate and profile. The same heat source and the same material of the model are used for the analysis in thermal condition for a fillet weld. This analysis performed by Barsoum et al. [5]

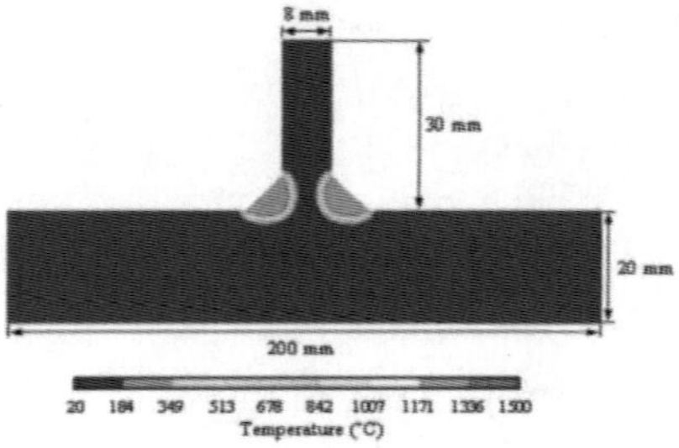

Figure 2: T-fillet welding performed to analysis thermal conditions in the thin plate. [5]

3.1 MATERIAL SELECTION

In the shipbuilding industry, two grade of steel is commonly used in the yacht building. First one is Grade "A" and the second one is Grade A36. Grade "A" Steel is most commonly used steel in the shipbuilding industry for the thin plate in yacht and ferry boats. This steel is mild and hot-rolled. Welding properties of this steel are excellent. [1]

3.2 COMPOSITION & PROPERTIES

Table 1: The chemical composition of Grade A steel. [6]

Element	Max. wt./ Content %
Fe	96.72
Mn	2.5
C	0.21
Si	0.5
P	0.035
S	0.035

Table 2: Thermal Properties of Grade A steel. [7]

Thermal Properties	Metric
Thermal Expansion	$11 \times 10^{-6}\,°C^{-1}$
Thermal Conductivity	$0.084 \frac{W}{mm \times °C}$
Specific Heat	$480 \frac{J}{Kg \times °C}$
Initial Temperature	20 °C
Final Temperature	1500 °C

5. MODEL

The dimension of the base plate is considered in these case studies are 1000mm in width, 1500mm in length and 5mm in thickness of the base plate. These dimensions are taken from the yacht building panel. The longitudinal flat bar dimensions are also taken from the yacht building panel, the dimension of the flat bar is 65x5 mm^2 and 1500mm long. The transverse flat bar dimensions are also taken from the yacht building panel, the dimension of the flat bar is 50 x 5 mm^2 and 450mm long.

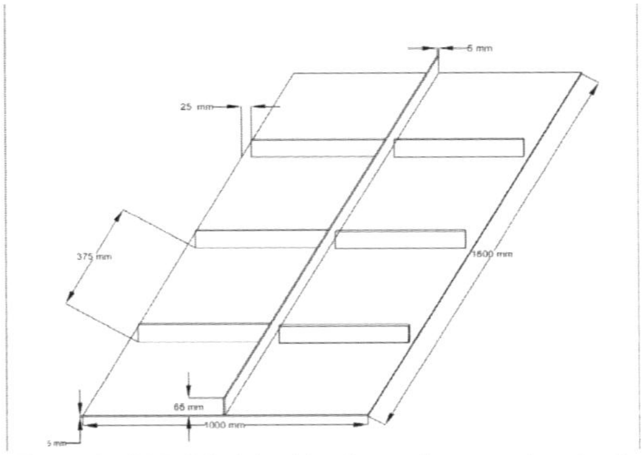

Figure 3: CAD Model with a base plate, one longitudinal flat bar and three transverse flat bars.

6. WELDING SEQUENCE

According to the thermal elastic plastic method, it explains the sequence of the welding which is used to reduce the deformation of the plate.

As we know that the sequence of the longitudinal frame:

First, we have to weld the stiffeners to the plate then the Frame is used to weld to the plate and at the end of the panel, we put the longitudinal Girder.

But from the different research by the scholar, they explain that if we change the sequence of the welding then we decrease the deformation of the plate.

They said that if we start from the girder and we can see the angular distortion of the plate. And as we know that if we want to reduce the deformation then we have to weld the plate and bar on the side of the girder.

So, we can weld the stiffeners to reduce the deformation that is angular distortion. After this, angular distortion is reduced. Then you can weld the frame to the panel plate to reduce the distortion more. [8]

The sequence of the welding this profile is shown below: [8]

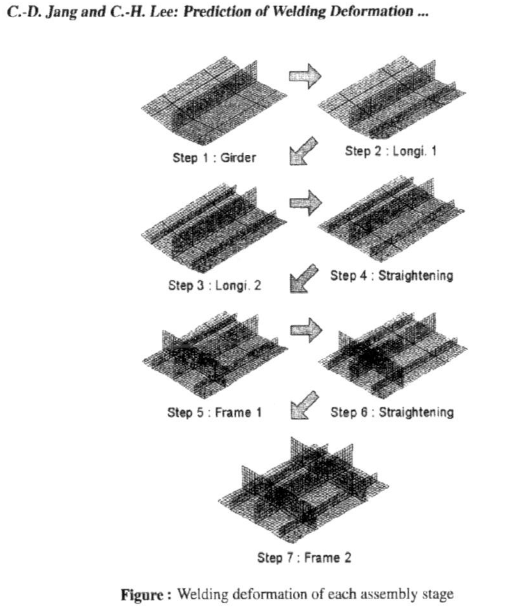

Figure 4: Welding deformation of each assembly stage. [8]

7. EXPERIMENTAL ANALYSIS & RESULT COMPARISON

Welding process and its complexities, it is required to perform a number of experiments to get the best possible result which is used to validate the analysis by the computational result. The same pattern of the computational method is used in the analysis for the large structure. Due to this reason, five case studies are performed which will represent in this paper. Out of that best possible result of one case study is used for the validation process. By this reason, different welding sequences from one to another case study or clamping time period are different in similar welding sequence. Different welding sequences are performed to reduce welding complexities. Complexity is occurred due to the temperature distribution in the model that is why different welding sequence is used to distribute the heat uniformly into the plate. It is another reason to perform different case studies.

In this method, uniform distribution effects of the thermal load affect the residual stress especially deformation. That is why five case studies are performed to analyse the best possible result of welding deformation. By comparing all case studies, one case study gives the best possible result of welding deformation.

For uniform thermal distribution, the thin plate is done through welding sequences. By this reason, different welding sequence is performed to reduce the welding deformation. These case studies are performed in the bulkhead panel of the ship.

In all five case studies different welding sequences are analysed which are given below:

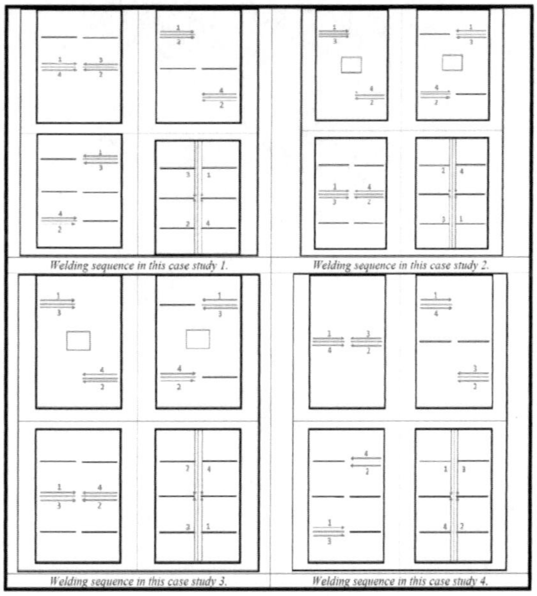

Figure 5: Welding sequences in the case study 1 to 4.

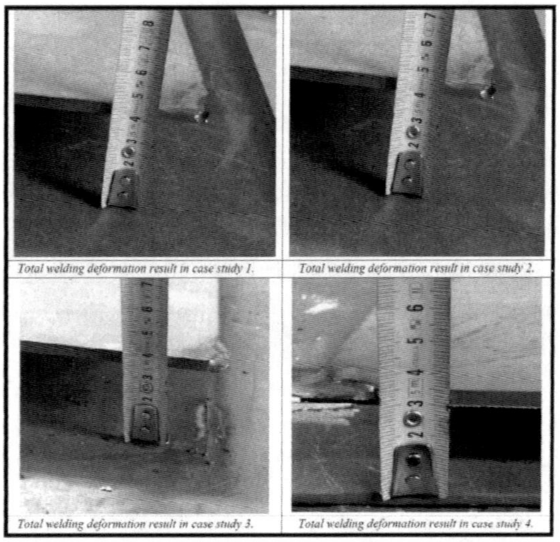

Figure 6: Total welding deformation result from case study 1 – 4.

Total welding deformation in the case study 1 is **43mm**.
Total welding deformation in the case study 2 is **42mm**.
Total welding deformation in the case study 3 is **37mm**.
Total welding deformation in the case study 4 is **30mm**.

Case Study # 5

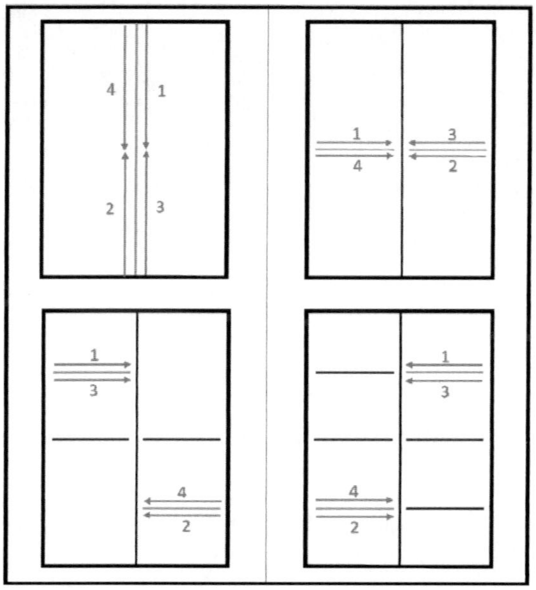

Figure 7: Welding sequence in the case study 5.

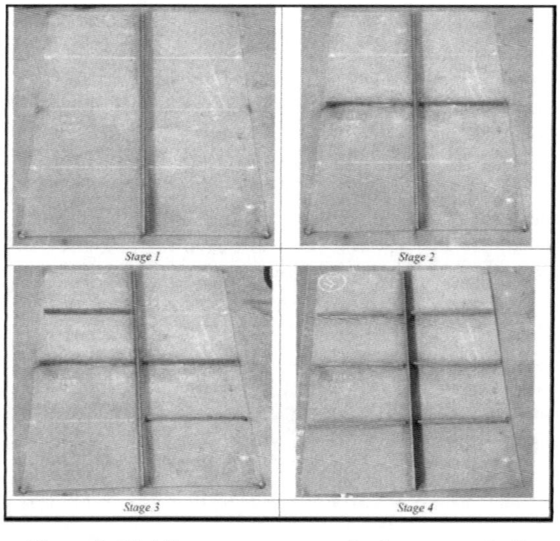

Figure 8: Welding process stages in the case study 5.

Total welding deformation in the case study 4 is **20.5 mm**.

This case study base plate is fixed from the corner of the plate during the welding process and after the welding, process plate is restraint up to 15 – 20 min after completion. Clamping method is essential to minimize the deformation in the thin plate.

After experimental analysis, it is proved that transverse plating method and time period of clamping during and after welding are increasing the productivity of the production.

8. VALIDATION

By the total deformation displacement results in the thin plate, welding sequence method is validated with the thermal elastic – plastic FEM method.

Experimental Result

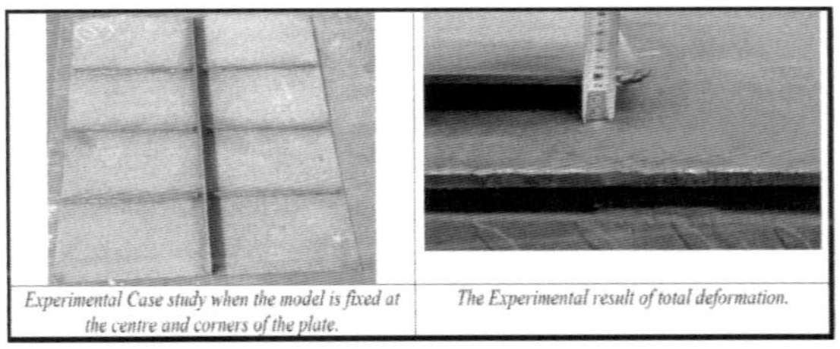

Figure 9: Total welding deformation result for case study 5.

In the experimental result, the maximum total welding deformation is 20.5 mm in case study 5, when the model is fixed from the corners and centre of the plate as shown in the above figure.

Computational Result

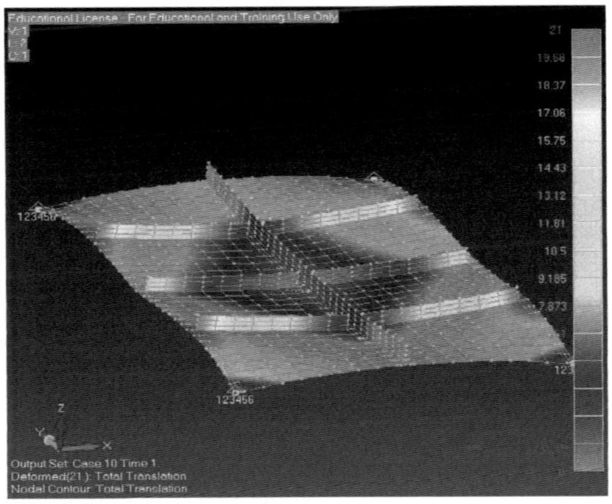

Figure 10: The computational result of total deformation. [1]

In this computational analysis, the number of nodes of the model is 1119 and number of elements are 1040.

In FEMAP this case study is to follow two steps to get von Mises stress or any mechanical physical parameters.

1st Step is used to analyse the thermal analysis i.e. temperature distribution in the plate and profiles.

And then 2nd step performs for the structural analysis by this analysis von Mises stress can be measured.

In the computational result, the maximum total welding deformation is 21 mm, when the model is analysed in the FEMAP software and the model is fixed from the corners and centre of the plate as shown in the above figure.

After experimental and computational analysis, the error between both the results is **2.4%**.

9. CONCLUSION

After validation of the experimental analysis and computational analysis, it is clearly shown that welding sequence in experimental analysis gives the good agreement of result with computational analysis which increases the productivity of the production process. The difference between computational analysis and experimental analysis is **2.4%** that one is too small that`s why it is neglected.

Through this welding sequence, **12-14%** of the production process of the bulkhead panel increases with respect to the labour cost and total time of production.

This welding sequence and computational results help to implement in the complex and large structures.

10. RECOMMENDATIONS & FUTURE WORK

To control the distortion during and after the welding process through:

- Eliminate the flaming fairing method control the distortion and use the inductive fairing method.
- Parallel heating method after completion of the welding process.
- Use the pre-machining concepts.

Techniques and methods can be analysed in future to control the welding deformation is "Influence of clamping in welding Deformation".

Other virtual fabrication technology models have formed to analysis the welding deformation in the thin plate.

By this method or technique, shipyard work on their own virtual simulation or develop their own code to analyse the ship structure.

11. ACKNOWLEDGEMENTS

This study has been accomplished in the framework of EMSHIP – European Masters Course in Advanced Ship and Offshore Design (www.EMSHIP.EU), coordinated by University of Liege, Belgium. I acknowledged the support by the Research Centre of Naval Architecture Faculty from University of Galati and Damen Shipyards Galati.

12. REFERENCES

1. ALI M. T., Computational Analysis to Control Welding Deformation in Thin Plate, *Design & Operation of Passenger Ships, RINA, 2019.*

2. From the slide of *Nirma University.*

3. PAZOOKI A.M.A, HERMANS M.J.M., RICHARDSON I.M., Control of welding distortion GMAW of AH36 plates by stress engineering, *Int J Adv Manuf Technol 88:1431457,* 2017.

4. JOHN LUCAS, Welding Procedure Specification's (WPS), *UTI Corporation.*

5. BARSOUM I. B. Z., Residual stress effects on fatigue life of welded structures using FEM, *Engineering failure analysis (no. 16) 449 – 467,* 2009.

6. www.shanghaimetal.com/shipbuilding_plate-129.htm

7. IACS Rule # 47.

8. C.D. JANG and C.H. LEE, Prediction of Welding Deformation of Ship Hull Blocks, *Journal of Ship & Ocean Technology (No.4) 44-49, 2003.*

13. AUTHORS BIOGRAPHY

Muhammad Taha Ali holds the current position of Research Intern at Damen Shipyards Galati. He is responsible for developing new methods to control deformation in the shipbuilding process.

ACOUSTIC ANALYSIS AND MEASUREMENTS OF MEGA YACHTS DURING CONSTRUCTION

M Insel, Hidroteknik Nautical Design Technologies Ltd, TURKEY
S Gokcay, and Z Saydam, Hidroteknik Nautical Design Technologies Ltd and Piri Reis University, TURKEY,
T Soyaslan and **C Soyaslan,** Soyaslan Design Ltd, TURKEY

SUMMARY

Noise is a critical parameter for super/mega yachts which can be verified only in the final stage of a mega yacht building project. Although there are more and more advanced methods to predict noise prior to the sea trials, verification has to be delayed until the noise survey is conducted during the sea trial. A new methodology is proposed based on measurements during construction to determine the transmission losses of both airborne and structure borne noise and propagation of the sound from the source to the receiver using these measurements. A 66 meter mega yacht case study is presented with measurements of airborne noise emitted through an omni-directional dodecahedron loudspeaker and measurements of structure borne noise generated by a tapping device. Both sound pressure levels and vibrations are measured to derive the transmission losses. A source-path-receiver method-based prediction tool, SNoPP, is employed to project the measurements into the final noise predictions. Comparisons between the measurements and the predictions are also presented.

NOMENCLATURE

ω	Frequency (Hz)
P	Pressure (N m^{-2})
RT	Reverberation time (s)
RT_{60}	Reverberation time for 60 dB drop
R	Room constant
S	Surface area (m^2)
L_P	Sound Pressure level (dB)
L_W	Sound power level (dB)
V	Volume (m^3)
α	Absorption coefficient
TL	Transmission loss (dB)
t	Thickness of material (m)
HVAC	Heating, ventilation, air condition
SNoPP	Shipboard Noise Prediction Program

1. INTRODUCTION

Ensuring acoustic comfort on-board mega yachts is one of the challenging tasks for yacht designers. At the design stage of a mega yacht, it is possible to estimate the final sound pressure levels in each cabin using source-path-receiver methods through air-borne, structure-borne and HVAC sound propagation. These methods rely on material properties of hull, internal division of cabins, furniture details, insulation material details and HVAC system characteristics.

Owing to the construction method of the mega yacht hull, three different types of material groups are added in different construction stages which are namely hull structure including decks and bulkheads, furniture and outfitting including non-structural insulation panels, and finally HVAC outfitting. Hence, the verification of predictions is normally performed after completion of all construction and outfitting work during the sea trials, after which it is very costly, both in financial and duration terms, to take corrective action if technically ever possible.

The currently proposed approach is to test the mega yacht at four different stages during construction to verify the predictions: after the structural hull construction, after the outfitting of all furniture either on shore or in the water, after completion of HVAC system, and finally during the sea trials. A newly developed yacht/ship onboard noise prediction program (SNoPP) based on source-path-receiver methodology is utilised at each stage to derive the absorption and transmission losses and to predict the noise levels at final sea trial condition. Although the first three stages do not change the necessity of the sea trial noise measurement verification, they can enable the designer/constructor to identify non-compliant cabins in terms of noise exposure budget at early stages in order to take precautions in a timely and less costly manner.

2. MODELLING MEGA YACHTS FOR NOISE ANALYSIS

Noise prediction techniques have been used over last 40 years based on source-path-receiver methodology [1-10] in three steps.

a) Noise sources are determined; their source strengths are identified either based on factory tests or by using previous case studies.
b) Transmission values between source and receiver compartment are evaluated for structure borne, airborne and HVAC transmission.
c) Receiver compartment direct and reverberant sound pressure level is calculated using absorption properties of the receiver compartment.

Although each step is crucial on the final noise levels, main uncertainty originates from the transmission mechanism and room reverberant characteristics. Noise transmission is through the structure and air, if HVAC transmission is accounted separately, which is crucial for the noise levels at anchor.

The sound sources such as propulsion engines emit acoustic pressure waves into the air surrounding the engine, i.e. airborne transmission mechanism. These longitudinal pressure waves reach the borders, i.e. decks or bulkheads of the source compartment and vibrate the border which in turn creates acoustic pressure waves into the neighbouring compartments.

Secondly, the engines or other sources create vibration on their foundations which can be described as acoustic waves advancing on the structural material. These waves travel through the structure reaching even the furthest compartments of the hull but somehow attenuated through the distance and structural arrangement of decks and bulkheads.

The transmission mechanism through air and structure is sketched in Figure 1. The noise levels can be compared with the IMO Criteria [11] or other criteria such as classification society comfort rules or owner requirements. Noise control options can be summarized as reduction of structure-borne noise through foundations/mountings [12], insulation applied to decks or bulkheads [13- 17].

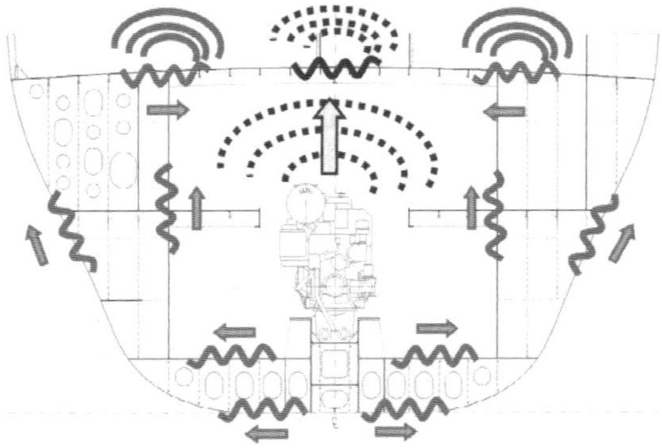

Figure 1. Transmission of noise through air and structure

The current case study is based on a 66 meter mega yacht, with steel hull and aluminium superstructure. A noise prediction analysis is carried out for the final noise levels, however the uncertainty due to transmission mechanism can normally not be verified before the sea trials, after which making any mitigation action is very costly if ever possible. Hence a sound transmission test programme was planned at four different stages:

a) During construction of the hull and superstructure before any noise insulation and furniture instalment,
b) After all insulation and furniture has been installed either on land or in water,
c) After HVAC system is fully installed and operational, HVAC noise can be measured independent of other noise sources,
d) Final validation during the sea trials.

Figure 2: Condition of the mega yacht during measurements

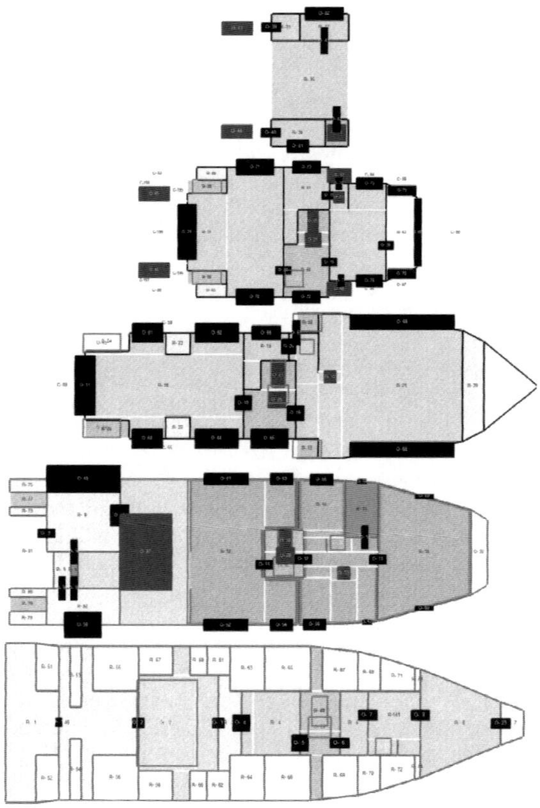

Figure 3: SNoPP Model for noise calculations

Currently, the first stage has been completed and outfitting of the vessel is ongoing. Figure 2 shows the status of the construction during the first set of measurements. In parallel to the measurements, a room-material model of the boat was created in Shipboard Noise Prediction Programme (SNoPP). SNoPP is based on division of ship volume into hexahedral volume elements. Each hexahedral element has material properties on its 6 faces. These material properties can be specified as combination of different layers. Openings can also be defined to model connections between the hexahedral elements, or openings to the outside such as windows. One or more hexahedral volumes can be combined to form rooms. The rooms are then combined to form compartments, which are surrounded by structural elements of the hull. Figure 3 demonstrates the model for each deck on test case including openings both vertically and horizontally.

Noise sources can be located into any hexahedral element and their airborne and structure borne sound power level can be described.

The task of predicting sound pressure levels in SNoPP is then reduced down to transmission of sound waves into different hexahedral volumes/compartments through both airborne and structure borne transmission. An extensive test campaign has been conducted to determine the material properties and five different validations of SNoPP prediction algorithms were made during the tests.

3. REVERBERATION

Sound pressure level in a room can be expressed as summation of three different components:

$L_P = L_{P-S} + L_{P-D} + L_{P-R}$, where:

$L_{P\text{-}S}$: Sound pressure level from direct sound source in room

$L_{P\text{-}D}$: Sound pressure level from boundary of the room

$L_{P\text{-}R}$: Sound pressure level from reverberant field

Reverberation component should be estimated from the room size, shape, materials covering the walls and materials used inside for furniture and furnishing.

Reverberant sound can be calculated from:

$$L_{P-R} = L_W - 10\lg(R) + 16$$

where

L_W : Total sound power level from direct sound

R : Room constant

Room constant can be calculated from

$$R = \sum_{i=1}^{6} S_i \alpha_i$$

where

S_i : Surface area for each material

α_i : Material absorption coefficient

Reverberation time is the time for the sound to die after the sound sources cease. Standard reverberation time is defined in terms of the drop in intensity from the original level, in general for 60 dB. The reverberation time can be estimated using Sabine equation as:

$$RT = \frac{0.161\,V}{R}$$

where

V : Room volume

Hence reverberation time measurements can be used as verification of room material absorption calculations, leading into more precise reverberation noise calculations.

A balloon was used to generate a sudden high intensity noise during the measurements. The change of sound pressure level was recorded in high frequency and 20 dB drop in the noise level was measured. 60 dB reverberation time (RT_{60}) was calculated by multiplying 3 and the time for 20 dB drop.

Figure 4 demonstrates a typical sound pressure change time history record when a balloon is burst. Such signal can be converted into dB decay form as given in Figure 5, which in turn can be described as reverberation time in each frequency band of 1/3 or 1/1 octave bands as given in Figure 6.

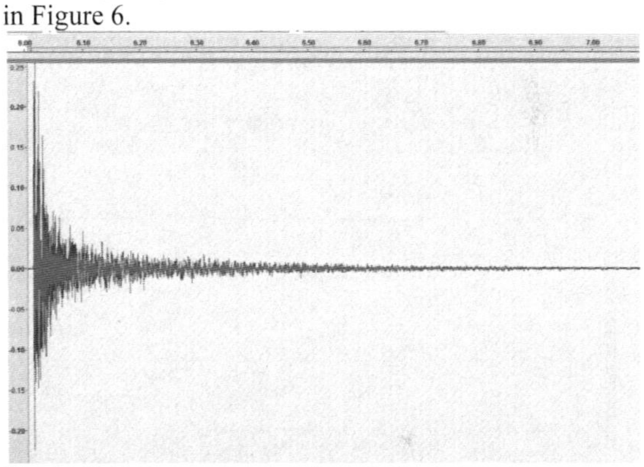

Figure 4: Pressure change after a balloon burst

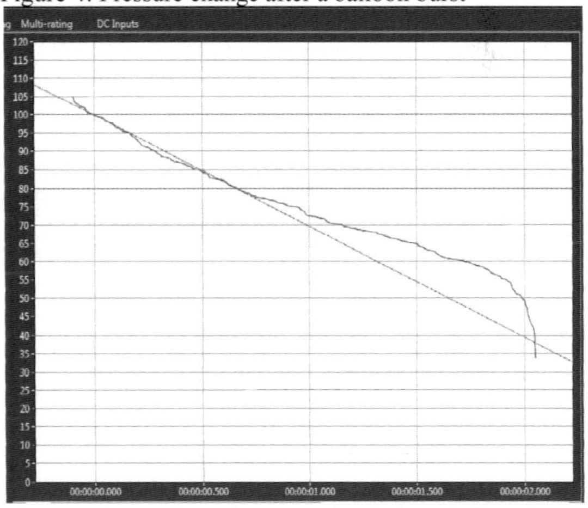

Figure 5: Drop of sound pressure level after a balloon burst

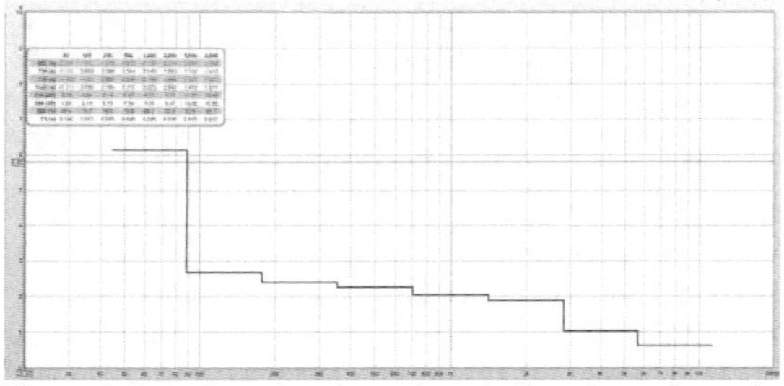

Figure 6: Reverberation time after a balloon burst for each octave band

Reverberation time shall fully depend on the materials on the surfaces of the compartment and interior furniture and furnishing. In the current case there is no outfitting, and whole absorption is generated by the main structural materials and openings between the compartments. Some compartments also exhibit limited insulation using mineral wool for heat insulation, tiled deck/bulkheads for sound insulation and pre-manufactured panels for division of rooms. Each compartment absorption is calculated according to the material content and compared with the measured values.

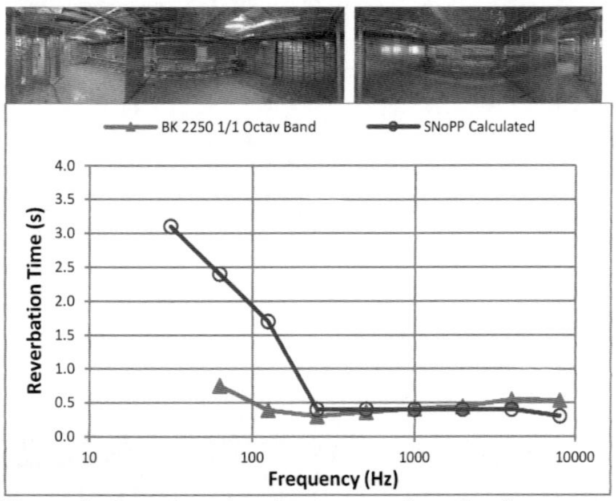

Figure 7: Reverberation time comparison for the master cabin area

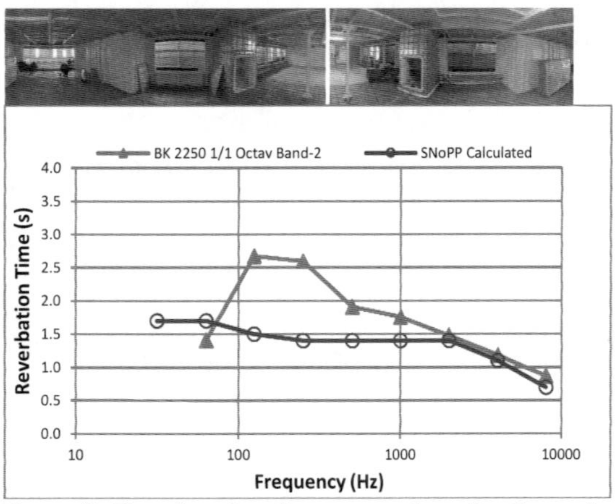

Figure 8: Reverberation time comparison for the main saloon

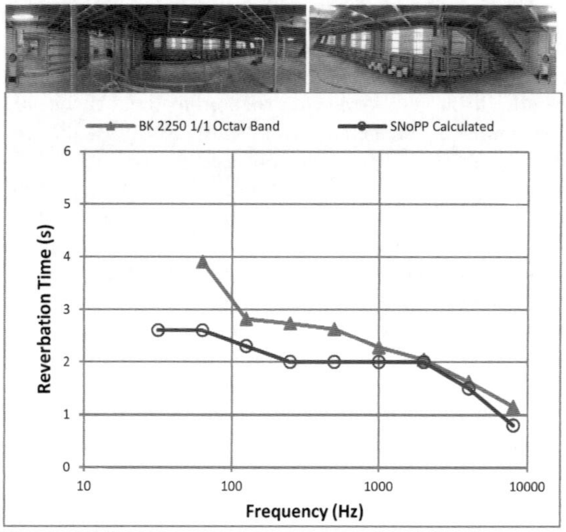

Figure 9: Reverberation time comparison for the guest cabin area

Three example compartments are given in Figures 7 to 9. The agreement with the measurement and calculations are very good in frequencies above 1 kHz, generally satisfactory for the frequencies between 250 Hz to 1 kHz. Below 250 Hz, difficulty to measure the reverberation time leads to higher levels of uncertainty; which is reflected on the results.

4. AIRBORNE NOISE TRANSMISSION

Airborne noise was generated using a dodecahedron loudspeaker in several compartments. The sound levels are measured in neighbouring compartments that are transmitted through bulkheads and deck structures. In the current case the doors and windows have not been installed yet, hence the opening through doors and windows also transmit the sound.

Figure 10: Dodecahedron loudspeaker airborne sound generation setup in crew cabin area.

Figure 11: Dodecahedron loudspeaker airborne sound generation setup in galley and engine room.

Figure 10 and 11 demonstrate the airborne dodecahedron sound generation setup in various compartments. Figure 12 and 13 demonstrate the measurement positions and airborne sound transmission both measured and calculated through SNoPP. The general agreement between measurements and predictions by SNoPP is satisfactory for frequencies above 100 Hz.

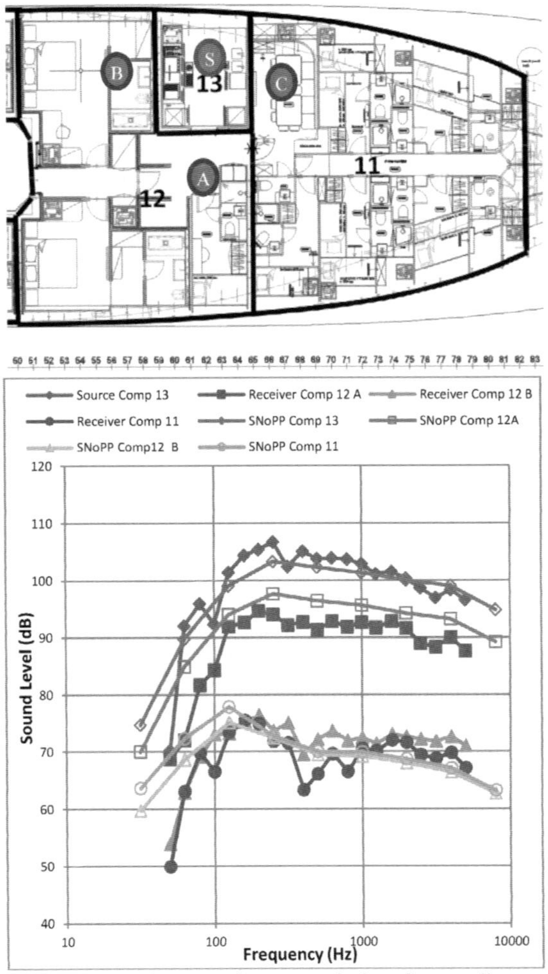

Figure 12: Airborne noise transmission from galley to neighbouring compartments.

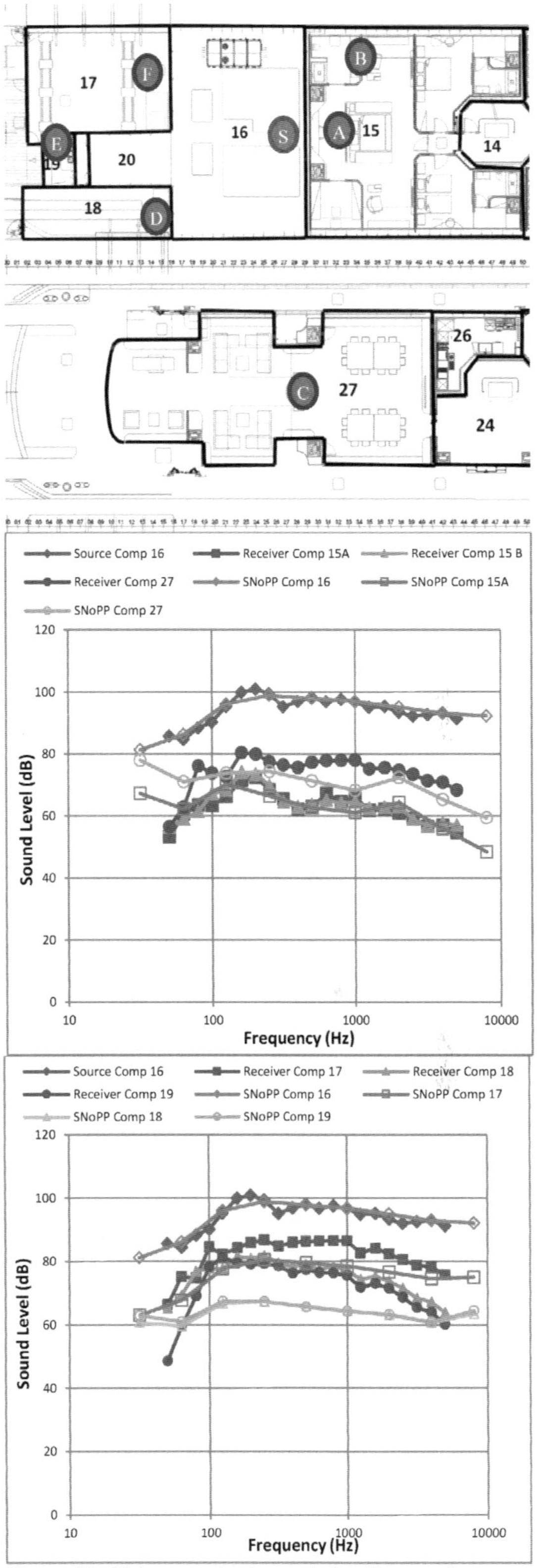

Figure 13: Airborne noise transmission from engine room to neighbouring compartments.

5. STRUCTUREBORNE NOISE TRANSMISSION

Energy from each equipment/source due to vibration excitation at the foundations shall be transmitted all over the ship hull.

Tests have been conducted on the yacht with a tapping machine producing impact on the application deck as shown in Figure 14. This type of impact source is particularly useful to test the noise transmission from upper decks to lower decks while walking in a finished yacht; directly influencing comfort of the passengers in mega yachts In the current case, a tapping machine has been used as vibration source in the source compartments with vibration excitation and propagation to neighbouring compartments has been assessed.

Structure borne noise transmission examples are given in Figure 15 and 16 from main saloon and engine room respectively, in the form of layout in top view and measurement &computation results. The attenuation of structure borne noise can easily be observed between points A-C-D in Figure 15. The distance between the source and point A, C and D is 0.75 m, 3.75 m and 14 m respectively. The reduction in structural vibration level is about 4.28 dB/m.

Figure 16 demonstrates the bulkhead reduction between B and C horizontally, as wells as between B, D and F vertically. The reduction in each + junction and 2.5 m steel bulkhead is about 20 dB.

Figure 14: Structure borne impulse noise generation by tapping machine in main saloon.

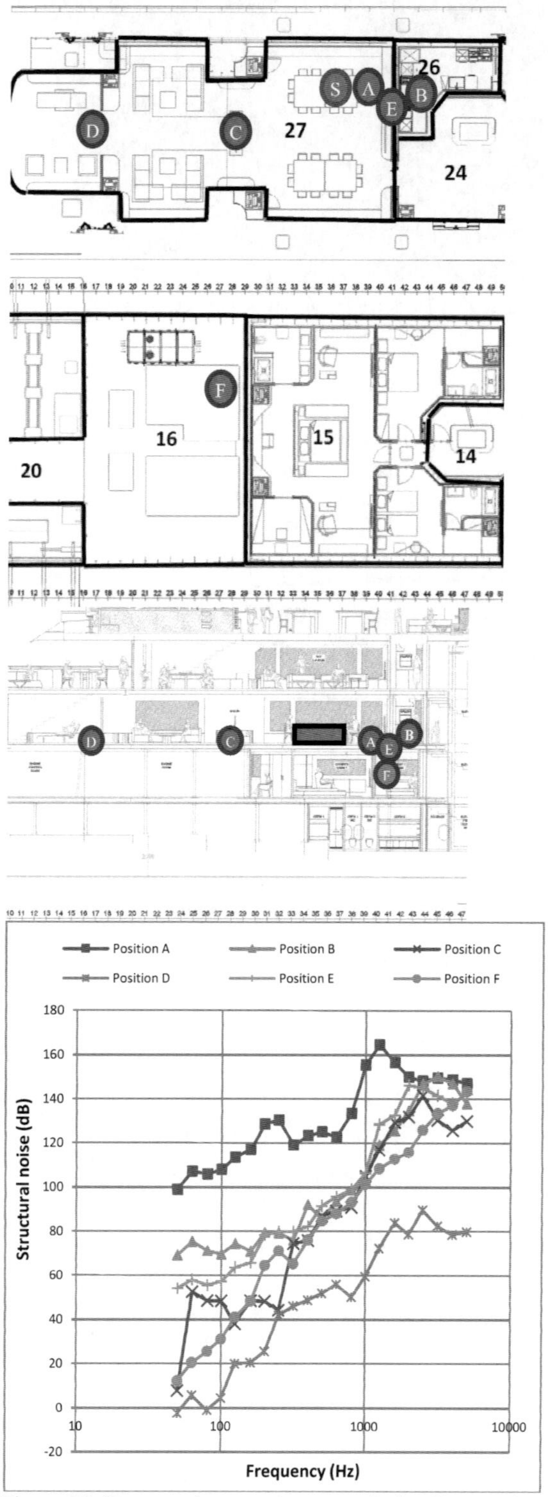

Figure 15: Structure-borne noise transmission from main saloon

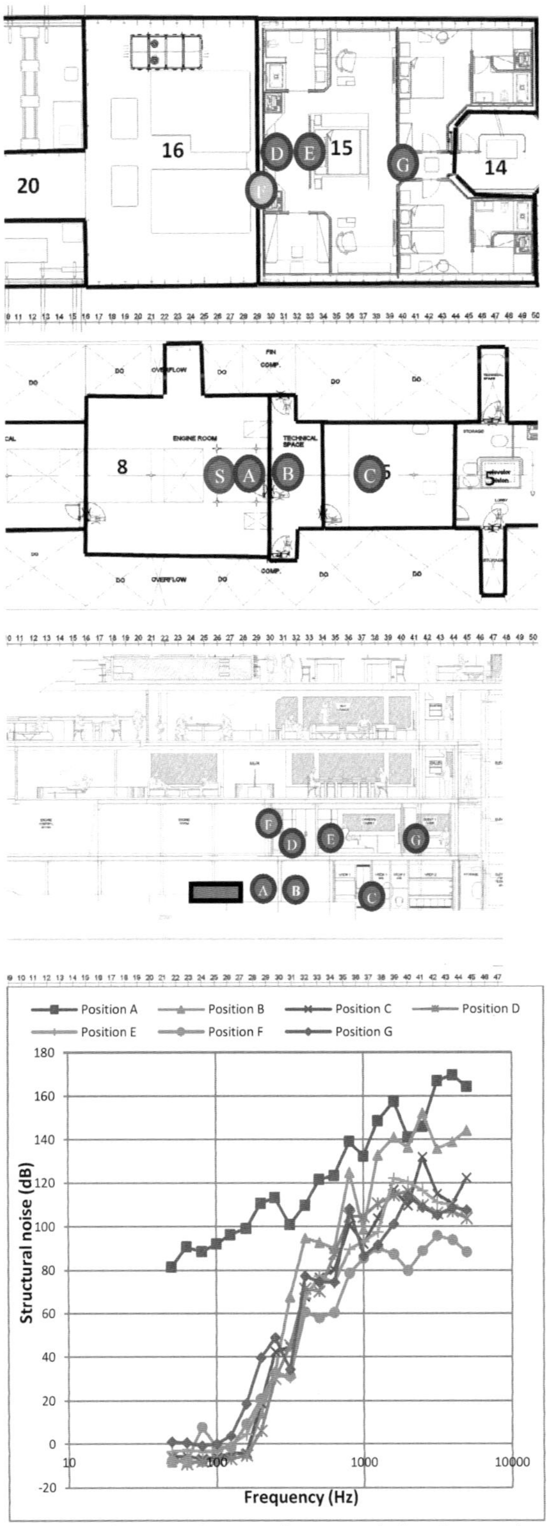

Figure 16: Structure-borne noise transmission from main engine room

6. AIRBORNE TO STRUCTUREBORNE NOISE TRANSMISSION

Airborne noise was generated in several compartments using dodecahedron loudspeaker, the vibration on the near structure was measured in all the three axes. This vibration shall be transmitted to the rest of the hull. The sound generation setup has been depicted in Figure 17.

Figure 17: Structural noise induced by airborne generated noise by dodecahedron loudspeaker in galley.

Figure 18 shows that significant structural noise is generated from the airborne sound signal reaching 80 dB at 500 Hz. The + junction between A and B results in a 10 dB reduction.

7 STRUCTUREBORNE TO AIRBORNE NOISE TRANSMISSION

The structure borne transmission also produces sound pressure changes in each compartment. In a compartment far from the tapping machine, noise is transmitted fully structure borne, however the decks/bulkheads create sound pressure waves in the receiver compartment. Measurements were made for the cases at which the tapping machine was located in different compartments, and airborne sound pressure levels were measured in various compartments.

As structure-borne noise gets transmitted long distances it becomes airborne as shown in Figure 20. The source at main deck main saloon generates considerable airborne noise at the foremost compartments 11 and 1 (Point D and G) around 50 dB.

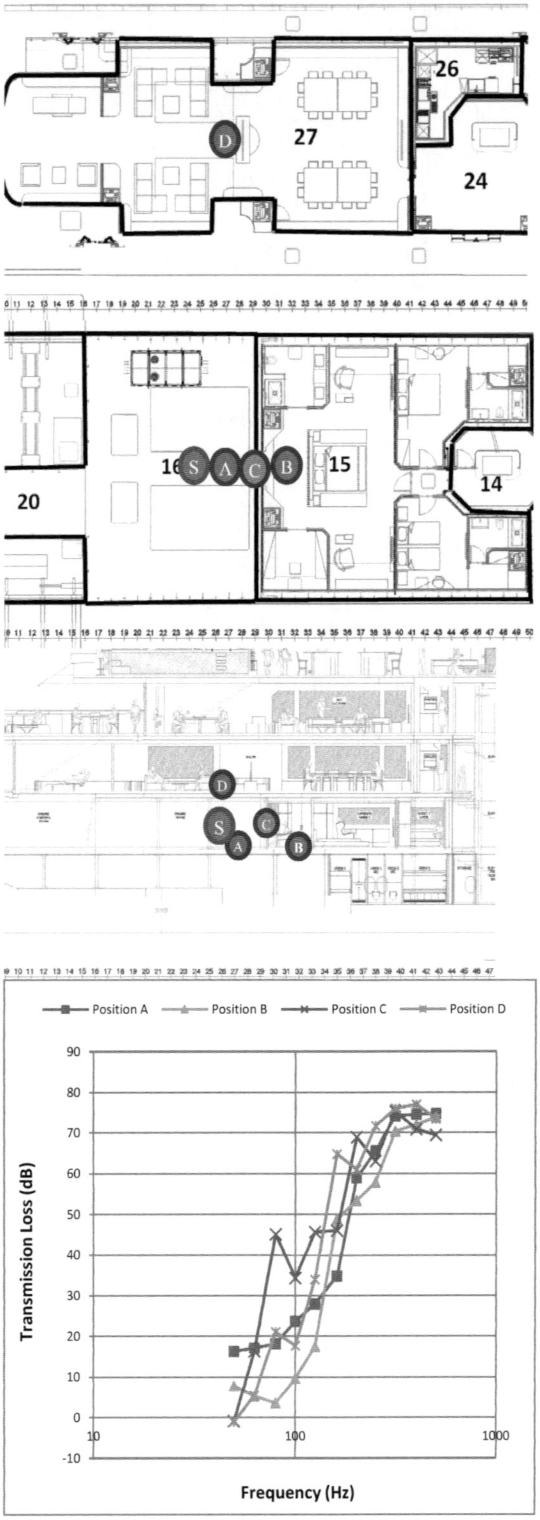

Figure 18: Structural noise induced by airborne generated noise by dodecahedron loudspeaker in engine room.

Figure 19: Structure-borne to airborne noise transmission measurements

8. CONCLUSIONS AND RECOMMENDATIONS

A shipboard noise level prediction program, SNoPP has been developed and a validation campaign has been started. Results of the first phase of the validation effort have been presented in this paper. With the proposed methodology, the reverberation characteristics of compartment may be accurately predicted for the majority of the frequency band; namely above 250 Hz. Airborne sound transmission predictions are in agreement with the measurements. Further developments are underway in order to better model openings during the hull construction phase.

It is possible to find airborne and structure-borne noise transmission losses using dodecahedron loudspeaker and tapping machine in an experimental fashion. The proposed methodology of conducting measurements during initial phases of construction -at the expense of approximately a day's long testing- for the prediction of final noise levels is found to be a plausible approach.

The second stage and third stage testing with interior furbished boat shall be conducted in order to obtain further information about room absorption characteristics and transmission losses to predict the final sea trial sound pressure levels through this methodology.

The tests conducted with the same methodology with a completed commercial ship has indicated that absorption coefficients and transmission losses can be derived from the current tests; leading to accurate ship trial predictions. The importance of second and third stage tests with a furbished mega yacht shall be very important as the variations of material absorption and transmission losses are considerably difficult to predict due to special furniture and materials used in mega yachts.

9. ACKNOWLEDGEMENTS

Authors would like to acknowledge the financial support from The Scientific and Technological Research Council of Turkey (TUBITAK), Ege Yacht and Mr Khaled Omayrat for allocating the mega yacht for testing.

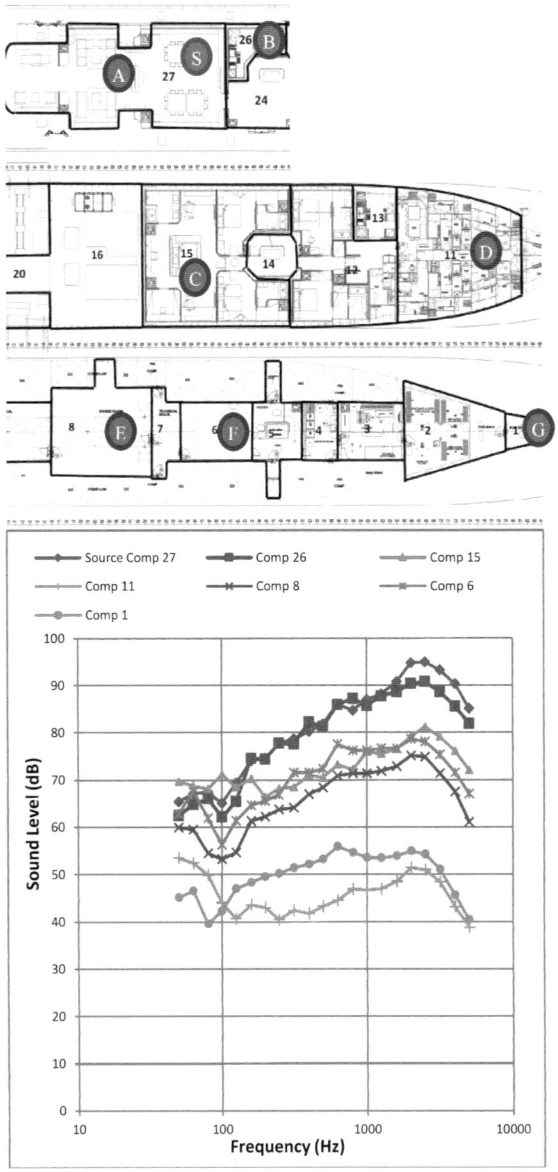

Figure 20: Structure borne impulse noise generation by tapping machine to airborne measured noise.

10. REFERENCES

1. NILSSON A.C., 'Noise Prediction and prevention in Ships', *Ship Vibration Symposium, 16-17 October 1978.*

2. FISCHER R.W., BURROUGHS C.B., NELSON D:L:, 'Design Guide for Shipboard Airborne Noise Control', *Technical and Research Bulletin N0:3-37, SNAME*, January 1983.

3. HOLLAND C.G., WONG S.F., 'Noise Prediction and Correlation with Full Scale Measurements in Ships', *Transactions of IMarE*, Vol 107, Part 3, 1995

4. INSEL M., UNSAN Y., TANG A.S.T., 'Noise and Vibration Considerations in Mega Yachts', *Proceedings of The Modern Yacht Conference*, RINA, 2007.

5. INSEL M., HELVACIOGLU I.H., UNSAN Y., 'Noise Predictions for Shipboard Accommodations', *Proceedings of 13th Congress of International Association of Mediterranean, IMAM*, Istanbul, Turkey, 2007.

6. FISCHER R., COLLIER R.D., 'Noise Prediction and Prevention on Ships', *Handbook of Noise and Vibration Control, 2007.*

7. WERK M., 'Ship On-Board Noise Propagation Analysis Methods', *Handbook of Noise and Vibration Control, 2007.*

8. SHURI H., 'Development of Noise Prediction Programs for Acoustic Ship Designs', *ClassNK Technical Bulletin, 2013.*

9. SHURI H., 'Development of Ship Noise Prediction System Using Vibration and Acoustic Data Base for Shipbuilder's Engineers', *Inter-Noise, Hamburg, 2016.*

10. 'Design Guide for Shipboard Airborne Noise Control', SNAME Technical and Research Bulletin No.3-37, 2018.

11. IMO, 'Code on Noise Levels on Board Ships', resolution *MSC.337(91)*, adopted on 30th November 2012.

12. FISHER R., BORODITSKY L., DEMPSEY R., JONES N., BAHTIARIAN M., 'Airborne Noise Flanking of Shipboard Vibration Isolation Systems', *Sound and Vibration, December*, 2006.

13. PETTERSEN J.W.E., STORM J.F. (Ed), 'Noise Control in Ships', *NTNF Report B 0930.4502.1*, Oslo, 1975.

14. VASSALO P., 'Experimental Study and Numerical Simulation of the Behaviour of Viscoelastic Materials to Reduce the Vibrations

on Board Ships', *PhD Thesis, University of Trieste, 2016.*

15. BIOT M., DE LORENZO F. 'Some Notes on the Sound Reduction Index of Pax Cabins Panels on Cruise Ships', Acoustics 08 Paris, *EuroNoise, 2008.*

16. INSEL M., HELVACIOGLU I.H., 'Effect of Noise Prediction, Treatment and Measurement of Ship Accommodations', *Proceedings of International Conference on Advances and Challenges in Marine Noise and Vibration, MARNAV 2012*, Glasgow, UK, 2012.

17. PIANA E.A., MARCHESINI A., 'How to Lower the Noise Level in the Owners's Cabin of a Yacht Through the Improvement of Bulkhead and Floor', *The 21st International Congress on Sound and Vibration, 13-17 July 2014.*

2. MIRENBERG K.J., 'Architectural Acoustic Modeling of Ship Noise and Sound Field Mapping', *Sound and Vibration, February* 2011.

11. AUTHORS BIOGRAPHY

Prof. Mustafa Insel (PhD) holds the current position of business development manager at Hidroteknik nautical design technologies Ltd. He is responsible for software development and physical testing for new applications. His previous experience includes noise and vibration measurements over 100 ships during sea trials.

Serhan Gokcay (PhD) holds the current position of technical manager at Hidroteknik nautical design technologies Ltd.. He is responsible for hull form development, design performance analysis and full scale testing. His previous experience includes testing of motor yachts, fast marine craft.

Ziya Saydam (PhD) holds the current position of general manager at Hidroteknik nautical design technologies Ltd. He is responsible for CFD studies. His previous experience includes hull form design of 20 motor vessels, and 5 sailing boats.

Turhan Soyaslan holds the current position of principle designer at Soyaslan design. He is responsible for mega yacht designs. His previous experience includes design of several motor mega yachts from 30 m to 70 m, and sailing yachts upto 40 m.

Can Soyaslan holds the current position of general manager at Soyaslan design. He is responsible for design analysis. His previous experience includes structural design and analysis of several mega yachts.

IDENTIFICATION OF TECHNOLOGIES AND PROCESSES TO ENHANCE THE SUSTAINABLE DESIGN, MANUFACTURE, OPERATION, AND END-OF-LIFE OF THE MOTOR YACHT ABOVE 24 METERS

M. Liang and **R. Birmingham,** Newcastle University, UK

SUMMARY

Sustainable development is a continuing concern in the Superyacht Industry due to the pressure on not only minimizing environmental impact but also considering economic viability and social acceptability. It is possible to enhance efficiency and effectiveness of resource use in manufacture, operation and recycle by identifying the sustainable technologies at the design stage. However, there is a lack of a holistic method to measure sustainable performance over the life cycle of the motor yacht during the technologies selection. A Life Cycle Sustainability Assessment (LCSA)based Sustainable Value Added (SVA) method has been developed to measure the sustainable performance for each technology using on the motor yacht over 24 meters. It is a new approach using the financial measure to aggregate data with existing economic terms to communicate with the investor with one language. This paper presents a systematic method and process to guide designer and owners to select and invest relatively sustainable solutions for the motor Superyacht.

NOMENCLATURE

LCSA	Life Cycle Sustainability Assessment
LCA	Life cycle assessment
LCC	Life Cycle Cost
SLCA	Social Life Cycle Assessment
LCIA	Life Cycle Impact Assessment
LCI	Life Cycle Inventory
SVA	Sustainable Value Added
NVA	Net Value Added
GVA	Green Value Added
MV	Monetary Valuation
RCR	Return to Cost Ratio
EU ETS	The EU emissions trading system
CBA	Cost-Benefit Analysis
GWP	Global Warming Potential
AP	Acidification Potential
EP	Eutrophication Potential
Co_2	Carbon dioxide
So_2	Sulphur dioxide
No_x	Nitrogen oxides

1. INTRODUCTION

Sustainable development is a global and long-term goal, and each of us on this planet needs to contribute. The owners' sense of responsibility for sustainability is increasing. It is necessary to consider sustainable design in the Superyacht industry to remain competitive. Moreover, motor yachts embracing inbuilt or delivered projects take up 87% of the superyacht projects in the world in 2017. Motor yacht is the major contributor to the finance, environmental and social issue in the Superyacht industry. To improve sustainable manufacture and operation over the life cycle, it is a necessary step for sustainable design to select applicable technologies applying on the motor Superyacht.

In this paper, a new approach is therefore developed to measure technologies monetary contributions to the three pillars of sustainability in life cycle perspective. It provides straightforward results on the allocation of finance, environmental and social resource for each technology. Moreover, it demonstrates how much more effective alternative technologies use resources compared with the benchmark technology. These overall results support decisions making during the technologies section.

The term LCSA was used to refer to 'Life Cycle Assessment (LCA) new' embracing Life Cycle Cost (LCC) and Social Life Cycle Assessment (LCA) as a supplementary impact in Life Cycle Assessment (LCIA). Therefore, LCA, LCC, and SLCA adopt the same goal and scope to define a single Life Cycle Inventory (LCI). LCA is known as environmental LCA which is a 'cradle-to grave' analysis method which provides an excellent understanding of the resource proportion. It helps to determine how changes in the process to avoid resource depletion. LCC is an economic assessment method as an accessory to LCA to ensure economic viability. SLCA is analogue to LCA but from a different perspective, e.g., such as health care, safety and so on. Therefore, the same framework applies to LCA and SLCA, yet different indicators are defined respectively [1]. There is a lack of transparent process and method to measure sustainability performance for products. In order to support the non-expert decision-makers in public in the real world, it is prerequisite to measure the efficiency and effectiveness of sustainable performance using LCSA result [2].

Sustainable Value Added (SVA) is a standardized tool to measure the company contribution of sustainability considering efficiency and the absolute level (effectiveness) of resource use. There are two measure ways to assess the company contribution to sustainability. One is the absolute measure to calculate Net Value Added (NVA) and Green Value Added (GVA) for one company at a time. Environmental and social damage is monetarized in the same unit so that it is possible to subtract between benefit and cost in the tree-pillars model. The other is a relative measure [3]. SVA is feasible for decision making, although it is widely applied to the company and never used on technologies selection.

Monetary Valuation (MV) in LCSA translates the environmental and social impacts into the monetary terms. Therefore, it is achievable to compare different environmental and social issues with the actual economic cost and benefit.

In this paper, a holistic life cycle thinking sustainable measure tool in the monetary term for technologies selection is proposed. A case study using hull materials is presented to verify the methodology. Technology selection will be a vital step for improving sustainable design. The LCSA result provides better understand fro the invisible cost and the resource flow in the LCSA system. The SVA results will provide an insight into the sustainable contribution of each technology and help decision making.

2. METHODOLOGY

The methodology proposed here quantifies the economic, environmental and social impacts. The aim is to develop an integrated system incorporating assessment, transformation, accounting and measurement for the decision-making process over technologies' lifespan. The configuration of the methodology includes four steps as illustrated in Figure 1. The first step is LCSA following the framework of ISO 14040 and ISO 14044. The main modification is adding LCC and SLCA in LCIA stage as the indicator of the economy and social respectively. The second step is to convert the environmental indicator and the social indicator into the monetary value. The results of the indicators are the cost of LCA and SLCA separately. The third step is the LCSA cost which aggregates the total cost of LCC, LCA, and SLCA. The purpose of this step is to show the internal cost and external cost of each technology over its lifespan. The final step is the SVA calculation and the results analysis. It supports decision making by understanding the sustainable performance of each technology for the motor Superyacht. The detailed method and process for each step are described in sections 2.1, 2.2, 2.3 and 2.4.

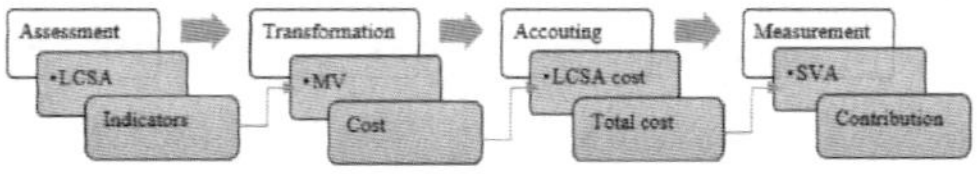

Figure 1: Holistic technology selection methodology

2.1 LIFE CYCLE SUSTAINABILITY ASSESSMENT

LCSA is an assessment tool to access economic, environmental and social impacts systematically. The method follows the standard of the ISO 14040 and ISO 14044. The framework of LCA includes four phases: goal and scope definition, Life Cycle Inventory Analysis (LCI), Life Cycle Impact Assessment (LCIA) and Life Cycle Interpretation (LCI) [5].

In this study, the subject is the technologies applied to the motor yacht over 24 meters over the 30-year life cycle. The assessment maps the allocation of the resources using on the technologies to identify sustainable technologies. The general system boundary included raw materials acquisition, inputs, and outputs in the leading manufacturing, energy supply, transportation, operation of products and end of life treatment. The functional unit data requirements depend on the technologies.

According to the goal and the scope, emission, energy and waste data of the technologies are collected and calculated. Data sources are from formula calculation, companies, journey article and technical data sheet of the products and other relative materials. The next phase is to select indicators. For sustainable assessment of the motor yacht technologies, the general indicators are shown as below.

- Economy indicators: LCC
- Environmental indicators: Carbon dioxide (Co_2), Sulphur dioxide (So_2) and Nitrogen oxides (No_x), emitted to air leading to Global Warming Potential (GWP), Acidification Potential (AP), and Eutrophication Potential (EP)
- Social indicators: health, safety, and comfort on board

Life cycle interpretation is the last phase of the assessment which reviews the scope of LCA and concludes the result of LCIA to carry out the MV.

2.2 MONETARY VALUATION IN LIFE CYCLE ASSESSMENT AND SOCIAL LIFE CYCLE ASSESSMENT

The purpose of monetary valuation is to transform the results of LCA and SLCA into the interchangeable unit to enable aggregate impacts with economic LCC over technologies' lifespan. MV is widely used in Cost-Benefit Analysis (CBA) to convert economic environmental and social impact into the monetary unit so that they enable to deduct with the financial cost or benefit. Massimo Pizzol suggests that market price method and hedonic pricing are used on environmental impact and budget constraint is applied to human wellbeing. Monetary valuation is the bridge to connect the three pillars of sustainability mutually [4].

2.3 LIFE CYCLE SUSTAINABILITY ASSESSMENT

LCSA cost is an accounting tool using the result from the monetary LCA and SLCA to estimate the LCC of the economic, environmental and social indicators. The process is based on the LCA system boundary and functional unit. The manufacturing cost is the initial investment. The data is from the result of LCC and the MV of LCA and SLCA. The operation cost and end-of-life treatment cost reflect the future price so that it needs to consider the time value of money due to inflation, so the formula as shown below.

Future cost = Current price x (1+ inflation rate) x t
(1)

2.4 SUSTAINABLE VALUE ADDED

Sustainable Value Added is a monetary measure tool. This study used SVA to measure the contributions of each technology to sustainability using available data. It considers both the efficiency and effectiveness of resource use — the effectiveness measures by CBA (Return to cost ratio). The purpose of a sustainable measure is to assess the contribution of sustainability including the three pillars of sustainability considering the efficiency and the effectiveness simultaneously [3]. The framework of sustainable value calculation includes three main steps: benchmark selection, Sustainable Value (SV) calculation, and results in interpretation [6].

2.4 (a) Benchmark selection

It is significant to define the benchmark for alternative technologies since the opportunity cost varies from the choice of different benchmark. A positive (negative) SV of the technologies addresses that similar technology has used the environmental source more (less) efficient than the benchmark technology [6]. The benchmark technology should be the most matured which means more researches to assess and more data to collect. The return figure is the LCC which represents the create return to the supplier. Therefore, the higher cost for the boat builder the higher return to the supplier. The data is collected from the result of LCC and the MV of LCA and SLCA.

2.4 (b) Sustainable Value calculations

The first step is to calculate the LCC of the technologies as a return created by the technologies.

Before calculating the opportunity cost, it is necessary to introduce eco- and social efficiency. This efficiency shows how much return created per unit of a resource [6]. Returning figure divided environmental or social impacts is equal to the eco- and social efficiency. In this study, LCC presents the return figure, and monetary LCA or SLCA are the environmental or social impacts. The formula is shown below.

$$\text{Eco} - \text{efficiency}(/\ \text{social efficiency}) = \frac{\text{LCC}}{\text{Monetary LCA}\ (/\text{SLCA})}$$
(2)

The next step is to calculate the opportunity cost. The opportunity cost is to measure the value added of the benchmark when it uses the same resource as the chosen technology. If selecting alternative technology, it will miss out the value created by using the benchmark technology [6].

$$\text{Opportunity cost} = LCA_{The\ chosen\ technology} \times Eco - efficiency_{benchmark}$$
(3)

The third step is to calculate the SV contribution using opportunity cost. The aim is to compare the value created by the chosen technology with the opportunity cost which is the value created by the benchmark technology [6]. The formula is shown below.

$$\text{Sustainable Value Contribution} = \text{LCC} - \text{Opportunity cost}$$
(4)

The final step is to calculate the Return to Cost ratio which is the conventional Benefit to Cost ratio which measures the sustainable efficiency of the technologies compared with the benchmark [6]. The formula is shown below.

$$\text{Return to Cost Ratio} = \frac{LCC}{\text{Opportunity cost}}$$

(5)

2.4 (c) Result interpretations

Sustainable Value Added (ASV) indicates sustainable performance from a financial perspective. The purpose of SVA is to show whose sustainable performance is better in financial terms, the chosen technology or the benchmark. The Return to Cost Ratio (RCR) is an indicator of sustainability efficiency [6]. The positive (negative) sustainable value means the chosen technology create more (less) sustainable value to the supplier but the higher (lower) cost to the boat builder and owner. If the RCR is greater than one, it means the resource using of the chosen technology is more effective than the benchmark.

3. CASE STUDY

This methodology was applied to the hull materials as a case study. The primary parameters, e.g., length, breadth, and draft, are collected from the same length motor yacht on the market. The rest of the settings are calculated according to ABS rules and regulations [7] and the books of Steel boat building [8]. There are five materials studied in this case study, wood, steel, aluminium, carbon fibre, and E-glass fibre. The function is to build the boat hull and use for 30 years. The particulars of the motor yacht are shown in *Table 1*:

Table 1 Particulars of the motor yacht

Ship particular	Steel plate	Aluminium 5086	Carbon-fibre	E-Glass fibre	Teak
Length [m]	24.98	24.98	24.98	24.98	24.98
Moulded Breadth [m]	5.50	5.50	5.50	5.50	5.50
Design Draft [m]	1.40	1.40	1.40	1.40	1.40
Moulded Depth [m]	3.42	3.42	3.42	3.42	3.42
Design Speed [Knot]	33.00	33.00	33.00	33.00	33.00
Hull net mass [Tonnes]	22.10	20.33	15.91	21.69	19.66
Displacement	56.30	51.80	40.54	55.26	50.08
Hull surface area [m2]	222.52	222.52	222.52	222.52	222.52
Welding distance [m]	1989.00	1829.88	0.00	0.00	0.00
Yealy fuel consumption(1000h)	433	397	315	421	385
Cutting area(10%) [m2]	22.25	22.25	0.00	0.00	0.00
Construction time [Hour]	4086	4495	2000	2000	6000

3.1 SYSTEM BOUNDARY

The unit processes are raw materials to a product, hull manufacture, operation and disposal as scrap. Transport is outside of the system. It assumes that the materials are from the local suppliers in China and the results of transportation are the same. Manufacture, maintenance and decommissioning of capital equipment are not included. It assumes that the initial investment for the machine is the same including all necessary equipment and tools for building different materials boat.

Moreover, other consumption and impact such as water, lighting, heating, are not included. It assumes that the building site and working environment are the same for all materials boat building. *Figure 3* presents the system boundary for the case study.

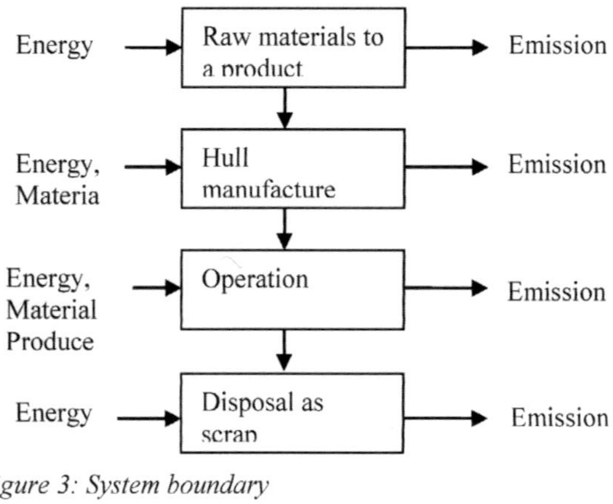

Figure 3: System boundary

3.2 ECONOMIC LIFE CYCLE COST

The LCC includes the materials cost, energy consumption cost, maintenance cost, the fuel consumption cost, and the scrap cost. In the metal hull maintenance process, it assumes the welding area is 5% of the whole welding area for each maintains. Carbon fibre and E-glass fibre hull maintain maintenance process is gel coat and repair. Wood maintains a maintenance process is vanishing and repair. In the final report of LCA-ship, it suggests 10% additional materials added over the life cycle of steel hull [9] [10]. It assumes the aluminium is the same maintenance process as steel Carbon- fibre, E-glass and wood are added 5% additional materials. The value of scrap materials presented as a positive impact in the LCA model. The value of scrap used the present market price. The price of raw materials was included in the material price. The results of the LCC cost are shown in *Table 2:*

Table 2: Economic Life Cycle Cost

Economic Life Cycle Cost		Steel plate	Aluminium 5086	Carbon-fibre	E-Glass fibre	Teak
Manufacture	Materials [Euro]	13790.40	40664.00	381888.00	28656.93	54576.60
	Zinc price [Euro]	3.88	3.88	0.39	0.39	0.39
	Gel coat [Euro]	0.00	0.00	0.00	0.00	0.00
	Raw material to product Electricity consumption [Euro]	31028.40	71693.07	809220.67	28201.37	158.46
	Welding Electricity consumption [Euro]	2166.82	2192.82	0.00	0.00	0.00
	Cutting Electricity consumption [Euro]	136.60	150.26	0.00	0.00	0.00
	Vacuum infusion Electricity consumption [Euro]	0.00	0.00	25236.43	34405.67	0.00
Operation	Welding Electricity consumption [Euro]	108.34	109.64	0.00	0.00	0.00
	Gel coat [Euro]	0.00	0.00	1134.00	1134.00	1764.00
	Zinc anode protection [Euro]	3.88	3.88	0.39	0.39	0.39
	Fuel consumption [Euro]	513499.01	470467.52	372929.45	499155.18	456123.68

3.3 ENVIRONMENTAL LIFE CYCLE COST

The monetary valuation was carried out in ISO 14008 standards, using the market price method to evaluate the environmental issue and social issue. Market price method defines the value for the lost product which is possible to use for resource depletion. The advantage of the methods is applied without considering uncertainty, but the disadvantage is the fluctuation due to the surplus and scarcity [4]. In this case study, the emission allowance price is used as the value of the emissions. The assessment model converted from different currency to Euro € for the base year of 2019 over the 30-year life cycle (2049). As is known to all, the three priority emission where are Carbon dioxide (Co_2), Sulphur dioxide (So_2) and Nitrogen oxides (No_x), emitted to air leading to Global Warming Potential (GWP), Acidification Potential (AP), and Eutrophication Potential (EP). The value of emissions, Carbon dioxide, Sulphur dioxide, and Nitrogen oxides, are identified by its emission market price. Emission market price is from The EU Emissions Trading System explained the EU emissions trading system (EU ETS) which is the world's first international emissions trading system, is aiming to reduce emissions by limiting emissions allowance and trade from the emission-saving projects around the world. It assumes the emissions in the research are exceeding the limit allowance and buy from the other emission-saving projects or companies. According to the report on EU's progress in cutting emissions in 2018, EU ETS announced 2.9% of emission reductions since 2013 which proves the system works [11]. The environmental indicator in this case study is the emissions from the process, energy consumption, and fuel consumption.

According to the world steel report, the primary energy demand of 1 kilogram steel plate production is 19.6MJ, and it generates 1.6 kilogram Co_2 emission, 0.0045 kilogram So_2 emission [12]. In the world aluminium report, electricity consumption is 13,562 kWh per tonnes of aluminium, and it generates 20 kilogram Co_2 emission per tonnes of o aluminium, 110 kilogram So_2 emission per tonnes of aluminium [13]. According to the carbon fibre company, ZOLTEK, the conventional raw materials of carbon fibre is polyacrylonitrile (PAN). Although carbon fibre is 35% lighter than Aluminium and 60% lighter than Steel, the Co_2 emissions of carbon fibre production are estimated to be 14 times than conventional steel production. The result of the research shows that pan carbon fibre production consumes 704 MJ per kilogram and release 31 kilogram Co_2 per kilogram [14]. According to the report of Life cycle assessment of CFGF – Continuous Filament Glass Fibre Products, 18.2 MJ per kilogram of primary energy are consumed to produce one kilogram of CFGF mat, and it generated 1.5 kilogram Co_2 emission, 3.6 gram So_2 emission and 0.9 kilogram No_x emission [6]. In this study, wood is from a renewable and sustainable source. It assumes the thickness of wood is 20mm. Research shows that energy demand is 26 kWh per cubic meter for producing 20mm timber [15]. In the OECD report, the carbon emission factor from electricity generation of China is 0.975 kilogram Co_2 emission per kilowatt hour.

For steel and aluminium hull, the primary process is cutting and welding. In this study, it included the energy consumption and the emission release for cutting and welding processes. For steel, Johnsen et al. report assumes that the cutting process consumes 8.5MJ electricity per square meter and loses 10% of steel. During the welding process, it assumes that welding distance is 90m per tonnes of steel based on the Johnsen et al. reported the net steel mass of a specific ship 1300 tonnes of steel and the welding distance is 117200m. It assumes the same ratio between welding distance and tonnage and 15.1MJ electricity per meter welding [16]. It assumes 10% greater electricity consumption than steel, for the cutting process remained the same as steel.

For carbon fibre and E-glass fibre hull, the main processes are layup and vacuum. In this study, it includes the energy consumption and the emission release for vacuum process but not included layup since layup has low energy consumption and emission release but high labour cost. The manufacturing method used the final process of vacuum infusion which energy consumption is around 22 MJ per kilogram of composite materials [17]. For wood hull, the construction method combines strip planking and cold mould which are low energy consumption and emission release but high labour cost so that it was not considered the energy consumption and emission release for wood. *Table 3* provides the breakdown of Environmental Life Cycle Cost.

Table 3: Environmental Life Cycle Cost

Enviromental Life Cyle Cost		Steel plate	Aluminium 5086	Carbon-fibre	E-Glass fiber	Teak
Raw material to product	Co_2 emission [Euro]	788.53	9.07	10999.97	725.64	0.00
	So_2 emission [Euro]	0.1850	4.1599	0.0000	0.1453	0.0000
	No_x emission [Euro]	0.00	0.00	0.00	273.34	0.00
	Electricity emission [Euro]	2593751.74	5995333.13	67671078.70	2358339.40	13251.29
Manufature	Cutting Electricity Co_2 emission[Euro]	1141.90	1256.57	0.00	0.00	0.00
	Welding Electricity Co_2 emission [Euro]	181130.33	183374.44	0.00	0.00	0.00
	Vacuum infusion Electricity Co_2 emission [Euro]	0.00	0.00	2110396.63	2877174.07	0.00
Operation	Maintain Welding Electricity Co_2 emission [tonnes]	77812552.13	179859993.98	0.00	0.00	0.00
	Fuel Co_2emission [Euro]	753738.55	690574.85	547403.79	732683.98	669520.28

3.4 SOCIAL LIFE CYCLE COST

Social Life Cycle Cost considers the salary of the workers. The length of the construction time indicates the difficulty level of the construction. The salary includes health insurance which indicates as the health of the workers. *Table* 4 presents the results of Social Life Cycle Cost in this case study.

Table 4: Social Life Cycle Cost

Social Life Cycle Cost		A, AH Steel plate	Aluminium 5086	Carbon-fibre	E-Glass fibre	Teak
Manufacture	Labor cost [Euro]	24516.00	26970.00	12000.00	12000.00	36000.00

3.5 MEASUREMENT

In this case study, steel was chosen as a benchmark. A positive (negative) SV indicates that the other materials have used its environmental and social resources more (less) efficiency than steel. *Table 5* shows an overview of the results of SVA and RCR for each material.

Table 5: The results of SVA and RCR

Valuation year	2019.00	2019.00	2019.00	2019.00	2019.00
Materials	Steel plate	Aluminium 5086	Carbon-fibre	E-Glass fibre	Teak
Manufacture Country	China	China	China	China	China
Trading Currency	EUR	EUR	EUR	EUR	EUR
Economic LCC [Million Euro]	21.80	20.00	16.90	21.20	19.40
Euro]	81.30	186.70	70.30	6.00	0.70
Social LCC [Million Euro]	0.02	0.03	0.01	0.01	0.04
Total cost [Million Euro]	103.20	206.80	87.20	27.20	20.10
Eco-efficiency	0.27	0.11	0.24	3.56	28.45
Social-efficiency	889.73	743.06	1404.90	1770.44	539.50
Benchmark Eco-efficiency	0.27	0.27	0.27	0.27	0.27
Benchmark Social efficiency	889.73	889.73	889.73	889.73	889.73
Euro]	21.80	20.00	15.60	21.20	19.40
Opptunity cost [Million Euro]	0.00	74.10	29.50	12.30	32.20
Sustainable value contribution [Million Euro]	0.00	(54.00)	(12.70)	9.00	(12.80)
Return to cost ratio	1	0.2700	0.5700	1.7300	0.6000

4. RESULTS AND DISCUSSION

The results of LCC show the total cost including environmental, economic and social cost of each of the technologies. The primary effect on the results is the environmental impacts cost. Over the life cycle, wood is the cheapest option, and aluminium is the highest option since the aluminium contributes to the highest environmental impacts. Wood contributes the least. The single assessment result indicates that the lowest cost for boat builder (highest return to the supplier) provides insufficient information for decision making since the results could be affected by one of the impacts. However, sustainable performance considers all three impacts, economic, environmental and social since the cost (return) created by the economic, environmental and social resource.

Therefore, a method is developed to communicate with the results from LCC. It provides a better explanation by taking steel and aluminium as an example. The return refers to the value created for the supplier which means the cost to the boat builder. Table 5 presents the summary statistic for SVA and LCC. In Table 5, aluminium generated 186.7 million Euro values of the environmental impacts over the life cycle. Meanwhile, it created the 20 million Euro economic returns for suppliers, so the eco-efficiency of aluminium is 0.11. Steel is the benchmark and used the same calculation. The results show steel generated 81.3 million Euro amounts of environmental impacts by creating 21.8 million Euro economic returns. The eco-efficiency of the benchmark is 0.27 which outperforms than aluminium in environmental resource use. If the benchmark products the same value environmental impacts as aluminium, it will create 50.4 million Euros which is 30.4 million Euros greater than the value created by aluminium which indicates the environmental contribution of aluminium is negative. 50.4 million Euro is the environmental opportunity cost of aluminium and -30.4 million Euros is the environmental value added compared with steel. The same calculation is applied to the alternative materials, and results reveal E-glass fibre created 9 million Euros more of sustainable value compared to steel which makes it the most sustainable materials in this study. The Sustainable Value Added to the supplier is placed in rank order which it is E-glass fibre, steel, carbon-fibre, teak, and aluminium.

Return to Cost Ratio is the indicator of sustainable efficiency. For instants, E-glass fibre created positive SVA which proves it has better sustainable performance than steel. In order to understand the efficiency of its' sustainable performance, it is essential to calculate Return to Cost Ratio. The Return to Cost Ratio of E-glass fibre is 1.73. It suggests E-glass fibre use its resource 1.73 times more efficient than steel.

From this study, knowledge gain is as follows:

- An understating of the resource allocation of each material
- From this information, relatively sustainable technology can be selected
- Insight for policy marker to identify and improve the problematic area of the technology over the life cycle

4. CONCLUSIONS

The method is a holistic and integrated method to assess the technologies of sustainable performance over the life cycle. It is possible to transfer environmental and social impacts in the same unit as economic cost and benefit. The case study proves the method can use for decision making on technologies selection for the motor yacht over 24 meters. There are few shareholders could benefit from this method.

The designer can use the method to select sustainable technologies and enhance the sustainable design of the super motor yacht.

It provides a straightforward indicator for the owner to gain insight into the sustainable performance of their Superyacht. □

The method can use for investment as a decision-support tool. Monetarized environmental impact and social impact will provide better information for new technology investment in a sustainable perspective.

Policymakers can use the result to identify the issue of the allocation resource in order to make more critical sustainable performance-based policy and regulation.

A further study will assess other technologies using the same method. The final sustainable design of the Superyacht will be proposed.

5. REFERENCES

1. Klöpffer, W., 'Life cycle sustainability assessment of products,' *International Journal of Life*, 2008.

2. Finkbeiner, M., Schau, E. M., Lehmann, A., & Traverso, M., 'Towards life cycle sustainability assessment,' *Sustainability, 2, 3309–3322.*, 2010

3. Figge, F. and T. Hahn, 'Sustainable Value Added—measuring corporate contributions to sustainability beyond eco-efficiency,' *Ecological Economics 48(2): 173-187.*, 2004

4. Pizzol, M., et al., 'Monetary valuation in Life Cycle Assessment: a review,' *Journal of Cleaner Production 86: 170-179.* , 2015

5. ISO, 'Environmental management —Life cycle assessment —Principles and framework,' *British Standard*, 2006

6. Frank F. Hahn. T, 'The ADVANCE Guide to Sustainable Value Calculations' *EU LIFE Environment programme*, 2006

7. American Bureau of Shipping, 'Guide for building and classing yacht, part 1', 2019

8. Thomas E. Colvin, 'Steel boat building-from plans to launching,' *Tiller Publishing*, 1996

9. Pommier, R., et al., 'Comparative environmental life cycle assessment of materials in wooden boat eco-design' *The International Journal of Life Cycle Assessment*, 2016

10. K.Jivén, et al., 'LCA-ship, Design tool for energy efficient ships – A Life Cycle Analysis Program for Ships,' *MariTermAB, TEM and SSPA, Sweden,* 2004

11. Progress made in cutting emissions, Available at https://ec.europa.eu/clima/policies/strategies/progress_en#tab-0-1, 2018

12. Brian Hughes William Hare Ltd World steel, 'LCI data for steel products,' *WorldSteel association*, 2012

13. World aluminium, 'Life cycle inventory data and environmental metrics for the primary aluminium industry,' 2017

14. Das, Sujit, 'Life Cycle Assessment of Carbon Fiber-Reinforced Polymer Composites,' *International Journal of Life Cycle Assessment*, 2011

15. Jingge Li, et al., 'Energy demand in wood processing plants', *Engineering*, 2006

16. Johnson T. et al., 'Screening Life Cycle Assessment of M/V Color festival,' 1999

17. John Summerscales, 'Composites Design and Manufacture', Available at: https://www.fose1.plymouth.ac.uk/sme/MATS347/MATS347A9%20NFETE.htm, 2006

6. AUTHORS BIOGRAPHY

Mincui Liang is currently an MPhil student in the School of Engineering at Newcastle University. The subject of her project is concerned with the life cycle sustainable performance of Superyacht. She has a background in Thermal Energy and Power Engineering. Her previous experience includes HVAC system design, ship design, building energy audit and financial analysis. These combined experiences and knowledge are used in this research to develop a holistic process to measure and decision making during technologies selection for Superyacht.

DEVELOPMENT OF HOVERCRAFT AS EXPLORER YACHT TENDERS

MJ Downer, Griffon Hoverwork Ltd, UK

SUMMARY

This paper discusses the use of hovercraft as tenders for Explorer Yachts to offer owners a unique vehicle with which to access locations from the sea inaccessible to conventional marine tenders. Further it considered the design considerations required in the mother vessel when a hovercraft is going to be specified as a tender.

1. INTRODUCTION

In recent years there has been an increasing interest from owners to cruise further afield than the traditional areas and explore more remote parts of the world. As a consequence, there is a growing market for new-build and conversion Explorer Yachts which are typically large with extended ranges, excellent sea-keeping, enhanced reliability and the ability to function for many weeks without support. When an owner decides that the thrill of this type of adventure is for them, they inevitably want to enhance their experience by getting closer to whatever scenery or nature they have sought out and to achieve this they require tenders on-board that offer this capability. Clearly relying on a limousine designed for use in the Mediterranean would be foolhardy when exploring an ice-flow or mangrove and therefore to compliment the 'go-anywhere' yacht owners are looking for 'go-anywhere' tenders. Robust RIBs offer an excellent choice for many situations but for some locations where rivers are in flood with submerged debris, beaches are unmade-up and ice sheets are too thick for breaking an amphibious hovercraft offers a unique additional opportunity. This paper discusses the approach Griffon Hoverwork have adopted to develop their hovercraft products for use as Explorer Yacht tenders and have overcome the unique integration issues presented with such a vessel.

2. HOVERCRAFT CAPABILITIES

2.1 PERFORMANCE

A hovercraft is a high performance marine vehicle which can operate at speeds in excess of 35knots in the right conditions. As they 'fly' just above the surface of the water they can be particularly affected by sea states especially short waves. The capability for a hovercraft to operate in seas which an Explorer Yacht owner may want to explore is driven predominately by length and skirt depth. Typically, at hovercraft with a length over 6.0m and a skirt depth of more than 0.5m will be most appropriate to operating as a tender for an Explorer Yacht and should be capable of a maximum speed in excess of 30knots up to a sea state 1 and over 20knots in sea state 2. These same hovercrafts will operate in higher sea states but at reduced speed usually below hump speed at less than 10knots. Smaller hovercraft particularly those with very small skirt depths are more suited to use as Superyacht toys in calm conditions within sheltered bays and riverine areas at Sea State 1. As the hovercraft has little contact with the water it is significantly affected by high winds and a typical limitation is 30knots gusting 35knots although it is noted that in these conditions exploring remote areas in a small marine vessel of any sorts is unlikely to be pleasant.

2.2 AREAS OF OPERATION

A hovercraft is a high speed marine vessel with the unique ability to operate over hard surfaces at similar speeds. In the context of the Explorer Yacht tender this offers the unrivalled ability to traverse onto and travel over:

- Very shallow water,
- unmade-up beach areas,
- frozen areas of water,
- mudflats,
- riverine areas in flood with underwater obstructions and rapids,
- marshlands,
- swamps.

Figure 1. Hovercraft operating on ice with mothership.

Griffon Hoverwork have an unrivalled number of over 180 hovercraft operating across the globe in 43 different countries. Typical uses for these hovercrafts include coastguard operations, search and rescue, scientific research, passenger transport, military operations, oil & gas operational support, airport rescue, humanitarian aid and environmental surveys.

2.3 DIMENSIONS & WEIGHTS

Hovercraft vary in size from 4.0m up to 35.0m in length and from 250kg to 115tonnes without payload. In the context of an Explorer Yacht tender able to operate in open seas safely in weather conditions that are likely to be encountered when an owner should reasonably expect to be able to use the craft the smallest practical hovercraft is 6.0m. Conversely the largest hovercraft practically able to operate from an Explorer Yacht garage at this time is approximately 8.5m in length. Later in this paper the possibility of future Explorer Yacht designs including a floodable well deck is discussed which would allow larger hovercraft to be used but for the purposes of this section smaller hovercraft able to be craned into a garage will be concentrated on. The latest Griffon Hovercraft design is the twin-engined, electric drive 995ED which has the following leading particulars when operating:

- Length 8.6m
- Breadth 5.2m
- Height 3.0m
- Cushion Depth 0.5m
- Seats: 8
- Max Payload 995kgs
- Max All Up Weight 3750kg

Like all hovercraft the 995ED has a flexible skirt that is inflated when hovering and therefore the size of the hovercraft when it is not operating is slightly smaller allowing it to fit into a smaller garage space. Additionally, the 995ED has inflatable flexible sidebodies which can be deflated in the garage to reduce the breadth even more. The dimensions for the same hovercraft when stowed in the garage are:

- Length 8.2m
- Breadth 2.2m
- Height 2.5m
- Lift weight for crane 2750kg

Due to the unique structural design of the 995ED it is possible to reduce the length of the hovercraft to approximately 7.0m with minor changes if only a smaller garage is available, but this does have a detrimental impact on performance and payload capability.

3. HOVERCRAFT TECHNOLOGY

3.1 STRUCTURAL DESIGN

Hovercraft hulls and superstructures are generally constructed either using light-weight marine aluminium or fibre reinforced plastic (FRP) in order to minimise their weight and maximise available payload. All Griffon Hoverwork hovercraft are designed to meet an appropriate international or classification standard specifically for hovercraft. These standards have been developed since the 1960's when commercial hovercraft was first being built and base their requirements on typical marine vessel loads plus wave impact loads experienced by hovercraft. In practice Griffon Hoverwork have found that an aluminium structure using 5083 sheet and 6082 sections offers the lightest weight and most robust solution especially when customers want the assurance offered by Classification Society rules and construction over-sight. A difficultly found when using such light scantling aluminium is the distortion and reduction in strength caused by arc-welding. In order to over-come these issues Griffon Hoverwork routinely uses Friction Stir Welded planking if this is possible, rivets structure above the waterline and in recent years has developed an adhesive bonding technique to eliminate welding in the structure of the 995ED.

3.2 MACHINERY SYSTEMS

Historically hovercraft was powered by gas turbine engines but Griffon Hoverwork pioneered the introduction of cheaper diesel engines into hovercraft. Today most hovercraft in the world use diesel engines with only the largest military hovercraft using gas turbines and some of the smallest unregulated hovercraft using petrol engines. Whichever type of engine is used in a fully amphibious hovercraft none of them will use seawater cooling and therefore the engines are derived from industrial, non-road or road engines. Because the engines used are not from the marine sector, they are inevitably certified for emissions control under a different scheme than MARPOL Annex VI. For the small hovercraft discussed here for tenders the engines should be below the current 130kW limit for NOx emissions. In order operate a hovercraft of this size it is essential to have separate control of the lift and thrust power and this can be achieved in a number of ways such as separate thrust and lift engines, variable pitch propellers or hydraulic or electric motors. In order to provide redundancy in the machinery system it is advantageous to have two engines both of which can provide thrust so that it is possible to reach a safe refuge or at least station keep if one engine fails.

Drive from the engine(s) to provide thrust and lift is transmitted by a number of methods depending on the hovercraft type, but the most successful systems use toothed drive belts and pulleys or more recently in the 995ED an electric drive allowing the option for hybridisation. Lift is provided underneath the skirt using one or more lift fans which depending on the size of hovercraft can either be axial or centrifugal; except in the most modern design of the 995ED which uses two mixed-flow fans. The trust of all fully amphibious hovercraft is produced by a propeller or two. In large hovercraft the propeller may be derived from an aircraft type and should be approved for use by an Aviation Authority; these are typically variable pitch to allow the thrust to be controlled by the operator and to provide reverse thrust. In smaller hovercraft it is normal to use a smaller thrust propeller which has been approved for use by the World Hovercraft Federation as these are less costly and easier to maintain although they do not offer reverse thrust unless the drive can be reversed.

3.3 INTEGRATED SYSTEMS

Hovercraft offer an excellent platform for integrating other systems such as electro-optical cameras for low-light and night-vision, navigation and communications systems. Additionally, Griffon Hoverwork have successfully integrated the use of drones with hovercraft which offer an excellent opportunity to explore further away from the hovercraft so as to select suitable routes for the hovercraft to transit or potentially to seek out natural and geographic things of interest.

4. HOVERCRAFT OPERATIONS

4.1 SAFETY CODES AND CERTIFICATION

The certification of hovercraft around the world is not consistent. Internationally the High-Speed Craft Code offers a set of standards for larger hovercraft, but this is not suitable for the size of hovercraft likely to be used for Explore Yacht tenders. As far as the author is aware only 2 countries have national safety codes specifically developed for hovercraft; the United Kingdom Hovercraft Code of Practise and the New Zealand Marine Rule 40F. A number of Classification Societies have developed hovercraft Rules some of which are considered to be the safety standard of choice in their home country such as the Russian River Federation and China Classification Society. Most other Classification Societies also have their own rules available for hovercraft. It is important to ensure that even if the hovercraft will be operated in the most remote areas of the world it is designed and built to a recognised safety standard and it should be certified for use to the regulations in place for the Flag Authority of the mother vessel. For the size of hovercraft likely to be used as a tender the UK Hovercraft Code of Practise is an excellent benchmark if no other suitable regulations are available.

4.2 OPERATIONAL TRAINING

Operating a hovercraft is unlike any other marine vehicle and therefore adequate training of the crew must be undertaken prior to taking command and refresher courses should be considered on a regular basis. Typically, a qualified commercial RIB helm will be able to be trained to safely 'fly' a hovercraft in approximately 25 hours. It is normal to undertake training on a specific model of hovercraft with the manufacturer and then be provided with a Type Operating Certificate from that manufacturer.

5. HOVERCRAFT INTERGRATION WITH THE MOTHERSHIP

The use of hovercraft, sometimes known as Air Cushioned Vehicles (ACVs) or even Landing Craft (Air Cushioned) (LCAC), is something that a number of the world's navies has been doing for many decades. In a military scenario hovercraft offer a unique capability to access locations out of the reach of conventional vessels and it is exactly this which is off interest to the Explorer Yacht owner. As such the learning from the years of military operations is relevant to the new challenge of integrating a hovercraft with an Explorer Yacht. Integration of the hovercraft with the mother vessel requires a number of considerations such as communications, launching and recovery and access for passengers.

5.1 GARAGE ACCESS ISSUES

A typical Explore Yacht garage and its integrated gantry cranes are similar to the 'boat' bays used to hold smaller hovercraft on some military ships. Hovercraft are provided with four permanent lifting points which can be used with vertical falls or single point lifting beam arrangements for a single crane. In the context of the Explore Yacht the garage door is generally considerably closer to the waterline than a military ship and an pair of extending gantry cranes able to operate over the water offers a perfect solution for loading and unloading the hovercraft. The centre of gravity of the hovercraft can be adjusted during operations to optimise trim and therefore the same systems can be used to ensure that the lift can be carried out safely even if there is a compromise in the location of the gantries to accommodate other vehicles in the same garage. The biggest difficultly with lading the hovercraft is that the skirt will retain some water if it is floated prior to recovery and this should be allowed to drain during the lift and suitable drainage should be provided in the garage. The skirt itself will fall down loosely during the lift and if the height of the garage opening is limited a method should be adopted to 'wrap' up the skirt segments to ensure they do not become entangled as the hovercraft is recovered; this can be achieved with simple netting if necessary. Once in the garage with a hovercraft such as the 995ED with its inflatable sidebodies it is possible to deflate these to improve space. Access to the hovercraft for crew and passengers can be via any suitable ladder or pontoon; the hovercraft floats stably with or without the engines running if no lift is applied.

5.2 THE FUTURE? YACHTS WITH WET WELL DECKS

Perhaps the ultimate mothership for hovercraft operations is a ship with a floodable dock usually at the stern with an internal 'beach.' In military terminology this type of ship was traditionally termed a Landing Ship, Dock (LSD) although recently more modern variant have been given the classification of a Landing Platform, Dock (LPD), the only significant difference being hanger facilities for aircraft maintenance in the latter. Whilst at this time the author is not aware of any military amphibious ships being converted for personal use or new build Explorer Yachts including these features in their specification this is an area of interest in the future. The design of this type of amphibious ship goes back to the LSDs developed in the Second World War by both the US and UK military well before hovercraft were ever developed for practical use however the successful integration of these two vessels stared in the 1970's. The key feature of the LSD is a floodable well deck with an opening stern gate and, to ensure the safe operation of the ship, a substantial ballasting system. In general LSDs are large ships and LPDs even large but in recent times more navies are seeking an amphibious and expeditionary capability and in keeping with budgets the consequence is that smaller LPDs are being designed and constructed. As an example, the Thai Navy operate a 141m LPD which is not a great deal larger than the biggest Explorer yachts. It is therefore not unreasonable to consider that at some point in the near future the features of the LPD will be adopted into a yacht in order to enhance the ease of interaction with amphibious tenders so as to provide even greater access to difficult to access areas of coastlines. It is not just hovercraft which will benefit from this approach as the same features will provide solutions for transporting road vehicles and expedition stores ashore via landing craft without the need for difficult craning operations.

8. CONCLUSIONS

As the interest in Explorer Yachts increases offering owners the opportunity to reach more and more remote areas of the world some are now also interested in ensuring the embarked tenders extend and enhance their voyage of discovery. The amphibious capabilities of a hovercraft to access ice, rivers, estuaries and beaches is unique and as such they are now being specified as requirements. In order to provide a safe experience a good quality sea-going hovercraft is necessary rather than a very small light-weight recreational hovercraft. A hovercraft of around 8m in length can prove to be the ideal vehicle but consideration must be given to its integration into the mothership from the outset.

9. AUTHORS BIOGRAPHY

Mark Downer holds the current position of Chief Engineer at Griffon Hoverwork Ltd. He is responsible for all technical engineering issues within the company and has overall responsibility and authority for the design of all hovercraft. He currently holds 4 patents for aspects of hovercraft design and was the lead in the development of the 995ED and 12000TD hovercraft. His previous experience includes 8 years' service in the Royal Navy, the design of naval ships at BAE Systems and managing the Plan Approval Office in London at Lloyds Register.

IMPACT OF GLASS DESIGN ON MODERN MEGA YACHTS - COMFORT AND STRENGTH ASPECTS

M Radon, S Semrau and O Doerk, DNV GL, Germany

SUMMARY

The trend towards large scale advanced glass solutions for the design of modern cruise vessels and mega yachts is accompanied by potential risks regarding both comfort as well as safety and reliability. The comfort related aspects are mainly focussed on unacceptable noise radiation excited by typical excitation frequencies of the vessel. Here as a possible solution to mitigate the potential comfort related risks a simulation-based procedure for noise and vibration assessment of glass windows is be described and its effects on the noise performance of the design explained. Possible countermeasures in case that the analyses reveal an unacceptable risk for violating the relevant noise limits are presented. In addition to the typical areas of attention for the structural impact of large and often coincident glass windows here focus is laid on the possible benefits and challenges of a load carrying window design. These are discussed based on practical experience and different case studies. Special attention is paid to the detail design of the glass bearing system.

1. INTRODUCTION

Over the last years glass has significantly gained importance as an element for the design of modern cruise vessels and mega yachts. Large movable doors, windows and walls made of glass enable transparent and clean designs with unique atmosphere. However, the trend towards large scale advanced glass solutions is accompanied by potential risks regarding both comfort as well as safety and reliability.

While the comfort related aspects are mainly focussed on noise radiation of the window itself excited by typical frequencies of the vessel, the safety and reliability related aspects are on the one hand related to the effects of coincident openings affecting the structural stiffness and on the other hand to a possible partial load carrying design of the glass itself.

Historically for the design of a vessel windows and doors were treated as an outfitting component and not as a structural part. Looking at the traditional sizes and arrangements of windows and doors in maritime applications there was no need to change this approach. The efforts related with a more detailed investigation and consideration of the effects of the windows and doors themselves on the structural behaviour were simply not to be justified by the to be expected gainings in terms of structural optimization and reliability. The same is true when it comes to the comfort related aspects e.g. noise radiation of standard size windows was not in focus as the eigenfrequencies were typically well above the relevant frequencies of the main excitation sources.

For modern designs showing large and unconventional glass elements this situation has basically changed. In many cases the impact of large and often adjacent openings on the stiffness and structural strength is requires special measures e.g. a load carrying design of the windows themselves. Alongside with the size increase, the eigenfrequencies of windows decrease, moving directly into the relevant excitation frequencies yielding to a considerable risk of unacceptable noise radiation. It goes without saying that at the same time the average expectations towards comfort increase often expressed by stricter acceptance limits for noise and vibrations.

2. DESIGN PRINCIPLES

The design of glass windows and doors for maritime applications is largely influenced by requirements of international standards and class Rules e.g. 1. and 2.. The original safety role of a window or door is to ensure the tightness of the vessel against acting sea pressure loads. Due to this the requirements for a glass window system i.e. the required minimum glass thickness is depended from the location of the window in the hull e.g. the height above the waterline and whether the window is positioned at front or back wall.

Mineral, thermally tempered glass is by far the most common material for maritime windows. Commonly glazing systems consists of one or several layers of glass bonded with polymers to ensure a good load transfer between the single sheets of the strength glass.

Additional functions of the glass, like e.g. good thermal insulation, specific colouring or reduced radiation transition, introduce additional glass sheets, often separated with an inert gas or air layer. These additional sheets usually do not play an important role for strength, however they make the supporting in frame more complex. Generally thin sheets of large dimensions have lower eigenfrequencies and are easily becoming a source of noise problems.

2.1 TYPICAL WINDOW FITTING SYSTEMS

There are two basic ways of fitting the glass into the structure typically used for maritime applications:

The first solution, historically longer existing, is to clamp the glass in a double side standard frame. The frame is overlapping the glass on its both sides. One side of the frame, the so-called glass retainer is usually fixed with a large number of screws. The glass retainer is compressing the elastic seals on both sides of the glass and by this forcing the glass in its position in the frame. Additional supporting elements guarantee the central positioning of the glass in the frame and prevent any contact with the metallic frame.

The second solution, a much younger system, allows to apply a metallic overlap of the glass only from the inside (on the low pressure side). The glazing compound is glued directly on the glass retainer, framing system or even straight onto the bulkhead.

The traditional clamped framed solution is usually more complicated in manufacturing but ensures a relatively easy way of removing / changing the glass in case of any need of repair. Such windows are usually welded into the hull structure. Their frame is considered from structural point of view as a flange on a hull cutout.

The glued system reduces usually the frame production and installing costs. In the extreme case such a glazing is glued directly on the cutout in the hull and no framing is required at all. In order to properly and reliably fulfil its main functions, such a window need to be installed under a precise gluing procedure, that is not always easy to follow in a typical yard environment. Some additional safety measures like e.g. mechanical retainers are also mandatory. The change of such a glass requires a more skilled team than in case of a traditional framed window. In some cases, when the window needs to ensure a fire tightness, the traditional framing presents clear advantages.

The most fundamental design principal for all glass window systems and solutions is to exclude any direct contact between the glass window and the surrounding ship structure made of steel or aluminium. This is to exclude impact or friction induced damages of the glass window that would subsequently lead to a damage or failure of the entire window system.

By standard the separation between the glass window and the surrounding ship structure is ensured by an elastic sealing system with a clearance at least allowing to compensate the maximum relative window opening deformations.

2.2 DIMENSIONING

The dimensioning of the glass is based on the requirements of the class societies. However, these are typically derived from standards like ISO 614, BSMA 24/25 or more recent ISO 21005. In the latest rule edition DNV GL 1. and 2. aligned the definition of the design pressures for windows with the ones for the surrounding steel structure.

Converting back the glass thickness requirement formulae will obtain in most of the cases a limiting stress of 40 MPa. A thermally tempered glass usually withstands about 4 times more, but as glass is a brittle material that does not provide any additional strength reserves beyond the first signs of the failure relatively high safety margins have to be foreseen. The safety margins must safely prevent the glass from any first failure sign, as this would be coincident with the ultimate collapse point.

3. STRUCTURAL STRENGTH IMPACT

Obviously large opening and window areas have a significant impact and influence on the structural strength behaviour of a design or ship respectively. This is mainly due to a decreased stiffness i.e. increased flexibility of the structural design and often accompanied by high stress concentration factors especially when applied in globally high loaded areas e.g. the ends of superstructures which are often subjected to high shear forces. In many cases the effects are not limited to the stiffness but simply driven by a lack of available load carrying material. Special attention has to be paid when adjacent (large) openings are located in areas which are subjected to significant simultaneously acting local and global loads of comparable magnitude.

In a conservative way glass windows or doors are considered to be non-load carrying within the standard strength analyses approaches. The resulting deformations of the window openings can be directly applied for the dimensioning or check of the window sealing clearances respectively.

While this approach showed its applicability for nearly all past and current projects it is recently more and more discussed to consider the possible stiffening and load carrying effects of (large) glass windows for future projects. Mainly two reasons can be named as driving forces for this. A reduction of the necessary efforts to ensure a safe and reliable design e.g. the arrangement of insert plates at window opening corners, so called "mickey-mouse" cut back corner solutions often in combination with high quality smooth grinding etc. and a simple optimization intention by saving structural weight. There might be also cases in the future where a structural design cannot be realized without the consideration of the load carrying effects of the glass windows.

A prerequisite for any consideration of load carrying effects of glass windows is of course a dedicated "load-carrying" window system design. In principle it is possible to consider these effects with a numerical finite-element based strength assessment but depending on the particular sealing system this may lead to a non-linear problem to be solved. The permissible loads i.e. deformations as well as stresses of the glass itself are key to be not overpassed. As the design principle to exclude any direct contact between the ship structure and the glass remains thoroughly valid smart load carrying sealing solutions have to be designed and applied. The arrangement of nylon pad supported sealings is a frequently raised solution idea in this respect.

Figure 1: Typical deformations between the glass and the surrounding steel structure at a common window arrangement (FE-model)

In order to gather information on the possible effects to be achieved by taking into account the load carrying effects of glass windows DNV GL has carried out some benchmark studies in this respect. The main objective of these studies was the generation of a sound technical information basis supporting the decision if the to be expected benefits justify the accompanying additional efforts. The studies were based on existing vessels.

The results obtained in the above mentioned studies are encouraging. The effectiveness of the stiffness of the glass is depending on several parameters. One of the most important parameters is the ratio between the glazed area and the remaining steel structure. It goes without saying that the more glass (larger windows) and the less remaining steel, the more pronounced is the stiffening effect of the glass. A typical deformation state between the glass and the surrounding steel structure of a large window arrangement is shown in figure 1.

Some selected results from current designs with large but not "extreme" big windows provided the following benefits which can be considered to be typical:

The stress distribution in the decks was found to be different, due to the higher effectiveness of the upper decks in the global bending. The longitudinal stress in the decks can be reduced by ~5% in the double bottom and 10-15% in the superstructure decks respectively.

The vessel's hull bending deformation was reduced by abt. 4%.

The stiffening effect of the glass can be even more interesting in cases of transverse acceleration conditions. Here again it is very much depending on the amount of remaining steel in relation to the size of fixed windows / glass walls. Reductions of deformation in a transverse bulkhead opening as shows in Figure 2 in a magnitude of 22% were observed.

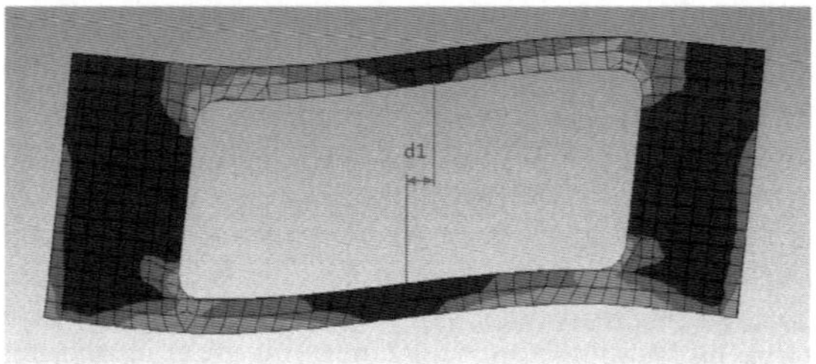

Figure 2: Deformations in an opening in a transverse bulkhead

As the ratio of glass vs steel can be expected continuing to grow, the future necessity to include the glass stiffness in the global behaviour of the structure seems to be obvious.

The inclusion of the glass stiffness in the global strength of the hull was also investigated in a R&D project led by the Technical University Hamburg-Harburg (TUHH) 3..

A notable positive effect regarding the reduction of the global deformation of the vessel was observed even with classic glued window solutions.

The window module itself was observed to increase the local stiffness of the frame and the surrounding structure even by eight times. However, the effect on the global stiffness is significantly smaller e.g. a reduction of the relative displacement between 2 decks of abt. 10% was detected.

The study focused also on the local effects and challenges related to the behaviour of the window frame system. Due to the constraining effect of the glued glass a different stress redistribution was detected leading to increased stresses in the corners of the window mullions. On the other hand, stress reduction could be observed in the outer corners of the frame.

For practical application of "load carrying" window designs the challenges related to different local stress distributions, the necessity to ensure the glass strength and buckling verification, or a definition of still permissible "damage" conditions for the structure when a glass is broken, will need to be addressed.

4. COMFORT IMPACT

As mentioned in the introduction, expectations towards comfort increases while limits for noise and vibration has been significantly decreased respectively. Typical, normal window dimensions in combination with class regulations are in general not critical regarding vessels self-induced global excitation frequencies. The requirements due to stability against sea wash and minimum glass thickness lead to relatively high eigenfrequencies which are well above the vessel's excitation frequencies.

The situation changes when the window dimensions starts to increase. The window eigenfrequencies are becoming lower and the risk of coincidence with vessel's excitation frequencies becomes more probable. Increased noise levels of up to 10dB due to noise radiation from windows are possible and have been observed onboard.

Eigenfrequencies cannot be avoided but shifted into more uncritical frequency ranges by window design modifications. Typical excitation sources for mega yachts and cruise vessels to be considered are the 1st and 2nd propeller blade passing frequency and up to the 1st and 2nd ignition frequency of the gensets or marine propulsion diesel engines.

To overcome this situation and provide a practical solution to minimize the related risks DNVGL established a lean service that allows the determination of the eigenfrequencies of large window systems in the early design stage as an addition to the standard noise prediction analysis for yachts and cruise ships. As changes during the design phase are more or less easy to implement and cost gentle it is of high commercial importance to gain the required information on potential noise limit violations already in an early design stage.

The procedure and the related service was developed together with a marine window manufacturer and has been successfully validated by several full-scale measurements. Figures 3 and 4 show photos of a full-scale validation measurement. The measurements confirmed the expected significant influences of the clamping grade of the surrounding structure, the sealing system as well as of the glass system itself including air gap, intermediate foil layers and their interactions.

Main goal in the development of this service was to generate a parametric finite-element based simulation model that can be easily adjusted but still include all necessary structures and couplings to calculate sufficient accurate eigenfrequencies within approximately 2 days. Figure 5 shows a typical parametrically generated finite-element model of a window noise analysis including deformations obtained from the analysis.

In case the simulation identifies a critical combination, counter measures are proposed to shift the eigenfrequencies into uncritical frequency ranges. Typical results of a window noise analysis are shown in Figure 6.

Typical counter measures are modifications at the support frames and the glass thickness.

Figure 3: Large scale noise radiation test for a window

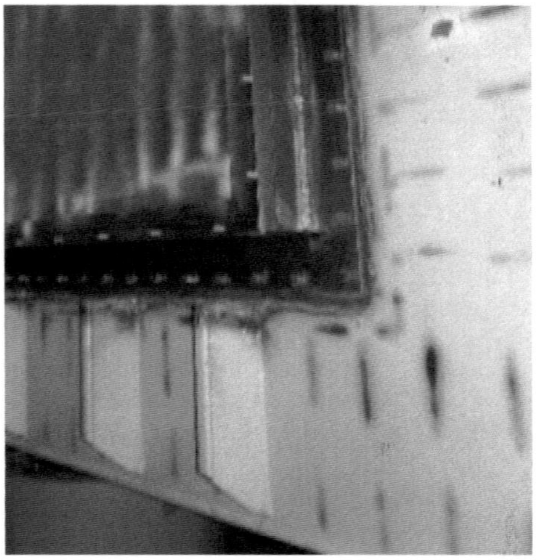

Figure 4: Detail view of the window fitting system

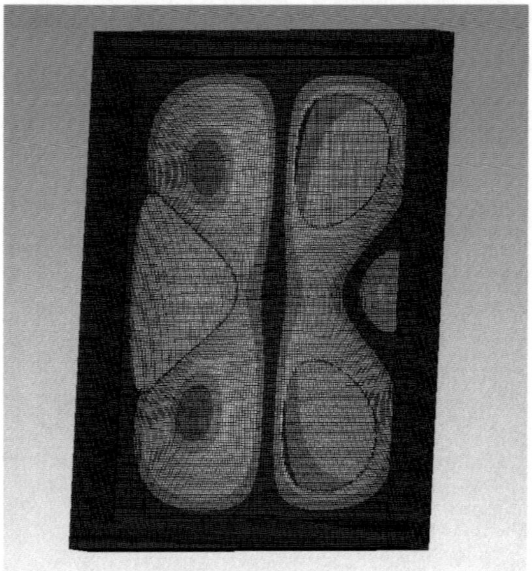

Figure 5: Typical finite-element model for a window noise analysis including deformations

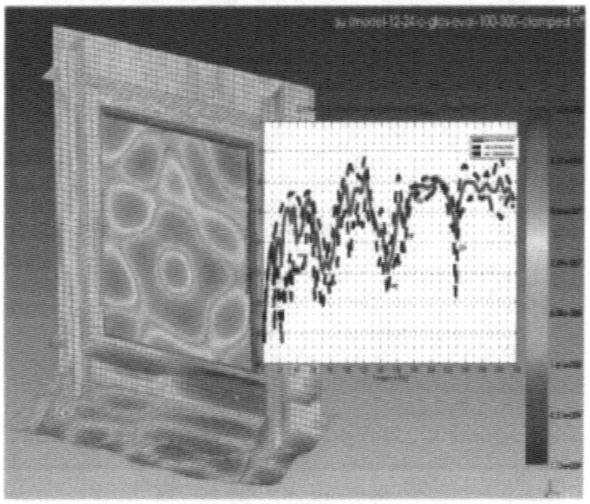

Figure 6: Typical results of a window noise analysis

4. CONCLUSIONS

Unconventional glass solutions including large glass areas will continue gaining more importance for the design of modern yachts and cruise ships. This development is accompanied by safety and reliability as well as comfort related challenges. Based on extensive R&D activities DNV GL has addressed these by generating advanced by practically applicable solutions and approaches.

Regarding the safety and reliability related impacts the consideration of a "load carrying" window design can be key to enable a design or can be applied in order to optimize a design in terms weight and complexity.
Consideration of load carrying effects can be recommended, for cases when cost and effort intensive structural solutions or modifications are unavoidable without consideration of this effect. Benefits can be expected for local as well as global deformations and stresses respectively. Experience is key to ensure a proper modelling, a sound sealing design and to avoid any overloading of the glass itself

When it comes to comfort aspects DNV GL's dedicated window noise analysis provides a sound and practical proven solution to eliminate noise radiation problems already in early design phase by application of relatively simple mitigation measures and thereby avoid any cost intensive late claim fix actions.

A window noise analysis is recommended for all window designs with more than 2.5m^2, especially if the window is located in the vicinity of propeller or the diesel engines. This is of course a rule of thumb and should be checked for each unique project.

Practical experience and especially the benchmarking with measurement results determines the quality of the analysis. Special attention must be paid on the glass system set-up including the air gaps and sealing system. This requires a dedicated modelling procedure to allow for an efficient but still accurate analysis.

So far, a very positive feedback was received from practical application of the window noise analysis i.e. no complaints were reported.

5. REFERENCES

1. DNV GL, 'DNV GL RU-Ship Pt.3 Ch. 12 Sec.6 Windows, side scuttles and skylights',2018.

2. DNV GL, 'DNVGL RU-Ship Pt.5 Ch.4 Sec.2.7 Glass Structures (Passenger vessels)',2018.

3. TUHH et. al., 'Einbeziehung geklebter Fensterbänder und Glasfronten in die Festigkeit von Schiffen'(in German), AiF funded research project, IGF project 18465 N, final report, 2017.

6. AUTHORS BIOGRAPHY

Dipl.-Ing. Maciej Radon holds the current position of Team Leader Strength and Fatigue at DNV-GL Maritime Advisory. He is responsible for design phase as well as various in operation phase related advanced calculations and investigations in the area of strength and fatigue. His previous experience includes the position of technical responsible for glass related products (windows / glass walls / glad doors / glazed sliding roofs) at Chantier's de l'Atlantique in Saint Nazaire.

Dipl.-Ing. Stefan Semrau holds the current position of principal engineer at DNV GL. He is responsible for DNV GL's noise related services like simulations, measurements and classification issues. His background is more than 23 years of practical experience in this field.

Dr.-Ing. Olaf Doerk holds the current position of Head of Department Structures at DNV GL maritime advisory services. He is responsible for the service offering of any kind of strength vibration and noise related simulation services. His experience and background is mainly related to strength, fatigue and fracture mechanics analyses of ship structures. This includes extensive R&D and committee work in these areas.

DESIGN FOR HELIDECK CERTIFICATION

C Wood, M Weetch and S Beech, Safeguard Helideck Certification, UK

SUMMARY

Consideration of the key requirements for a helideck in the early stages of design will ensure that customer the wishes and technical requirements can be met, without unduly constraining an innovative design process. It is the objective of this paper therefore to provide the Superyacht designer with some basic guidance to the design of a helideck, such that it can be demonstrated to be "safe by design" as part of this design process. This in turn may be used to support safe helicopter operations and provide a good basis for certification should it be required at any point during the yacht's life-time.

1. INTRODUCTION

The integration of one or more helidecks within the design of modern Superyachts is becoming a more common occurrence. Whether a helideck is provided for personal use, ad-hoc operational activities or for the use of charter customers the provision of a certified helideck that has been demonstrated to be "safe by design" is a key element to ensuring the safety and efficiency of helicopter operations.

This paper aims to look at how the integration of design codes (such as the Red Ensign Group (REG) Superyacht Code) during the design process can ensure a deck that meets the designers, the owners and certification requirements. The paper will consider some of the key features of a helideck relating to safe helicopter operations, with respect to certification, and will consider some of the lessons learnt from recent certification activities. As a Maritime and Coastguard Agency (MCA) Aviation Inspection Body (AIB), it is Safeguard's objective to provide certification without providing unreasonable restriction to an innovative design process.

2. HISTORY

30 years ago, a typical Superyacht was less than 40m in length. In 1985 motor yacht Buckpasser, at 36.8m in length, was one of the earliest yachts to boast a fully functional Helicopter Landing Area. Today's Superyachts are anywhere from 40m to 130m. This growth in size has allowed designers more space to incorporate helidecks into designs. Advances in materials, manufacturing techniques, design innovation and customer demands have led to a vast variety of technical solutions being built into current designs, and future projects. However, whatever design is chosen, thought needs to be included within the design process to allow a solution that can be the basis for safe helicopter operations.

The increase in helicopter use in support of Superyacht activities has led to an increase in associated incidents; of note and most recently in May 2017, the ditching of an Airbus A S 350 helicopter with three persons on board following a loss of control whilst trying to land on a yacht in the Port of Bergen, Hordaland county, Norway [Ref. 1]. Such events act as a timely reminder that design and operational considerations need to form part of the design of any yacht intended to carry out such operations, as the first step to ensuring safe helicopter operations.

3. RELEVANT ORGANISATIONS

The regulations associated with the design of helidecks are overseen by several organisations as follows.

3.1 INTERNATIONAL CIVIL AVIATION ORGANIZATION

The International Civil Aviation Organization (ICAO) is a specialised agency of the United Nations (UN) which works with 192 Member States. It codifies the principles and techniques of international air navigation and fosters the planning and development of international air transport to ensure safe and orderly growth. They also work with industry groups to reach consensus on international civil aviation Standards And Recommended Practices (SARP) and policies to support safe, efficient, secure, economically sustainable responsible civil aviation sector. SARP relating to offshore and shipboard helidecks are detailed within ICAO Annex 14 – Aerodromes Vol II – Heliports [Ref. 2].

3.2 THE CIVIL AVIATION AUTHORITY

The Civil Aviation Authority (CAA) is the statutory corporation which oversees and regulates all aspects of civil aviation in the United Kingdom. The organisation directly or indirectly regulates all aspects of aviation in the UK. In some aspects of aviation, it is the primary regulator, which includes helicopter operations to North Sea Oil Platforms. The CAA also produce civil aviation publications. Key to this paper is the CAA document Civil Aviation Publication (CAP) 437; Standards for Offshore Helicopter Landing Areas [Ref. 3] has become an accepted world-wide reference for offshore helidecks.

3.3 RED ENSIGN GROUP

The Red Ensign Group (REG) is a group of British Shipping Registers. It is made up from the United Kingdom, the Crown Dependencies (Isle of Man, Guernsey and Jersey) and the UK Overseas Territories (Anguilla, Bermuda, British Virgin Islands, Cayman Islands, Falkland Islands, Gibraltar, Montserrat, St Helena and the Turks & Caicos Islands) which operate shipping registers from their jurisdiction. The purpose of the Red Ensign Group Yacht Code is to provide design criteria, construction standards and other safety measures for yachts. Annex H of the Red Ensign Group Yacht Code [Ref. 4] outlines the minimum standards for Helicopter Landing Areas. This document refers out to ICAO Annex 14 – Aerodromes Vol II – Heliports [Ref. 2] regarding General Requirements of "Purpose-built Shipboard Heliports"; and "Purpose-built Shipboard Heliports provided in the bow or stern of the ship.

4 SAFETY FUNDERMENTALS

Helicopter operations onboard are becoming an integral part of the superyacht's function and as such these operations are also coming under greater scrutiny from design, commercial and regulatory bodies. As part of this greater scrutiny the drive for the demonstration of safety will increase. Fundamental to the demonstration of safety of any processor equipment are two elements;

- The equipment must be "safe by design";
- The equipment must be "used safely".

4.1 "SAFE BY DESIGN"

The equipment must be "safe by design". This is achieved by designing the equipment to meet technical requirements specified by the class and regulatory bodies and addressing any hazards that may be foreseen within the equipment's use at the design stage. Demonstration that the equipment is "safe by design" is achieved through the Certification Process which is considered later in this article.

The REG Yacht Code [Ref. 4], for example, presents two opportunities for the Certification of a helideck. The first is during the design process. This allows the designers to work with the certification body and presents the significant opportunity of ensuring the design meets requirements whilst it is still on the computer screen. This will significantly reduce the risk of changes being required to the built vessel, and of incurring the associated time and cost penalties. The second certification will initially be conducted on the built vessel at its commissioning, and then through life, to demonstrate that the helideck remains compliant with the technical requirements. This allows for changes in the superyacht structure or arrangements to be recognised and reviewed to ensure they do not affect the safety of helideck operations.

The Certification Process itself will be conducted against class and technical requirements such as the REG Yacht Code, as well as any specific requirements placed on the vessel by local authorities or proposed operations.

4.2 "USE SAFELY"

The equipment must be "used safely". This requires the training of all personnel and crew involved in the helicopter operations, ensuring that they are suitably qualified and experienced to use the equipment as the design intended. It should be noted that a good design will place less pressure on a busy crew to make up for the shortfalls in that design.

5 CONSIDERATION IN DESIGN

The design for a yacht intended to carry out helicopter operations should consider the following elements. In each case it is extremely important designers gain advice as early as possible to avoid difficulties or costly alterations later. As background to the design, the designers should establish at the outset what type of helicopter the owner wants and how it will be used, as that drives the design and the required equipment. For example, does the owner want to do ocean transits with his helicopter, which may mean a hangar is required? Does he need to fly at night, which requires a lighting plan? How much flying is intended and does the helicopter require onboard fuel or maintenance capability?

There are two approaches to a helicopter landing area. The first is the non-commercial or non-certified approach. This is where an owner will operate their Yacht and/or Helicopter purely for their own purposes and not carry or transport fee paying passengers. There are no set standards for such operations, and certification is not required. However, it is strongly recommended that the design requirements are still considered to support safe helicopter operations, and to support any change of use to commercial operations, as considered in the second approach.

The second approach is called fully certified or commercially certified. For superyachts flying the Red Ensign or its equivalent, particularly those in charter or that intend to charter, the primary regulations governing their design, construction and operation are found in the REG Yacht Code [Ref. 4].

5.1 HELIDECK LOCATION

In considering the location of a helideck, designers should look to how this may affect the operations of a helicopter. A helideck on the foredeck limits the capability, as the yacht will have to be stationary for recovery and take offs. Helidecks aft will have a wider operational envelope, but the effects of turbulence generated by the superstructure is greater and more troublesome. However, features may be designed and modelled to minimise the effects of airflow disruption.

5.2 DECK LOADING

All helipads, both certified and non-certified should be designed for a dynamic impact load that is 2.5 times the helicopter's Maximum Take Off Mass (MTOM) [Ref. 3]. Normally, decks are designed only for sea and human loading, and occasionally for point loads from tender stowage, but when helicopter operations are added, they must be strengthened, with added structure sometimes required from the helideck down to the keel. This adds weight, which increases overall displacement, adversely affecting speed and fuel consumption. Except on the largest yachts, helidecks are usually at or near the top of the yacht. This added weight, at such a distance above the keel, may have a significant negative effect on the yacht's stability, so added compensation features may be required within the design

5.3 D VALUE

'D-Value' means the largest overall dimension of the helicopter when rotors are turning. This dimension shall normally be measured from the most forward position of the main rotor tip path plane to the most rearward position of the tail rotor tip path plane (or the most rearward extension of the fuselage in the case of Fenestron or Notar tails).

On a certified helideck, the calculated D-value will be detailed within the certification documentation and will limit the size of helicopter that can operate from the helideck within the certification regime.

It is important for owners, designers and yacht management companies to consider what helicopter type they plan to operate to their helideck. This will dictate the minimum size of deck required to allow this. All Helicopter D-values are listed in the Reg Yacht Code Annex H [Ref. 4].

5.4 SURFACE FRICTION

The limiting friction values cited in CAP 437 Standards for Offshore Helicopter Landing Areas [Ref. 3] are related to the threshold at which a helicopter would be expected to slide on a helideck. The test method should therefore result in friction values that are representative of the 'real' situation of a helicopter on a helideck. For example, this suggests the use of a tester employing a tyre material to conduct the test.

Within certain limits, a helicopter could land almost anywhere on a helideck and an adequate level of friction should exist wherever the helicopter wheels touchdown. A measurement method is therefore specified that requires measurement of surface friction across the whole area of the helideck. Devices that provide automatic electronic data collection, storage and analysis are therefore to be preferred. The Helideck surface should ideally be dry but may be tested wet provided that any puddles are removed (e.g. swept into the guttering) and prevented from reforming during the survey.

Section of Helideck	Fixed Helideck	Moving Helideck
Inside TD/PM Circle	0.6	0.65
TD/PM Circle and H Painted Markings	0.6	0.65
Outside TD/PM Circle	0.5	0.5

Table 1 - Coefficient of Friction Requirements

To complete a full friction test the entire landing area is to be covered by parallel runs at no greater than 1m centres, repeated with orthogonal runs as shown in Figure 1, at the end of this paper.

5.5 OBSTACLES

An obstacle, as defined in ICAO Annex 14 [Ref. 2], can be a fixed (whether temporary or permanent) or mobile with parts thereof that:

- are located on an area intended for the surface movement of aircraft; or
- extend above a defined surface intended to protect aircraft in flight; or
- stand outside those defined surfaces and that have been assessed as being a hazard to air navigation.

It is advisable to limit the number of obstacles that may present hazards to helicopters within the Helicopter Landing Area. This will reduce the likelihood of a helicopter fouling its wheels or skid during take-off and landing. Examples of potential hazards include nets or raised fittings on the deck that may induce dynamic rollover of the helicopter.

The following definitions are given to the two Obstacle sectors associated with Helicopter Landing Areas, as illustrated in Figure 2, at the end of this paper:

- The Limited Obstacle Sector (LOS) - The 150° sector within which obstacles may be permitted, provided the height of the obstacles is limited.
- Obstacle Free Sector (OFS) - The 210° sector, extending outwards to a distance that will allow for an unobstructed departure path appropriate to the helicopter that the landing area is intended

to serve, within which no obstacles above helideck level are permitted.

Objects whose function requires them to be located within the Touchdown and Lift-Off Area (TLOF) (such as lighting or nets) shall not exceed a height of 2.5 cm. Such objects shall only be present if they do not represent a hazard to helicopters. Note that the TLOF and Final Approach and Take-Off (FATO) for a shipboard Helidecks are presumed to be coincidental.

For a TLOF of 1 D and larger, within the 150-degree limited obstacle surface/sector out to a distance of 0.12 D measured from the point of origin of the limited obstacle sector, objects shall not exceed a height of 25 cm above the TLOF. Beyond that arc, out to an overall distance of a further 0.21 D measured from the end of the first sector, the limited obstacle surface rises at a rate of one unit vertically for each two units horizontally originating at a height 0.05 D above the level of the TLOF.

5.6 EMERGENCY ARRANGEMENTS

The following fire-fighting appliances shall be provided [Ref. 4] in close proximity to the helideck and stored near the means of access to that helideck:

- at least two dry powder extinguishers having a total capacity of not less than 45 kg;
- carbon dioxide extinguishers of a total capacity of not less than 18 kg or equivalent;
- a suitable foam application system consisting of monitors or foam making branch pipes or Deck Integrated Pop-up Nozzles (DIFFS) capable of delivering foam to all parts of the helideck in all weather conditions in which helicopters can operate and which shall be capable of delivering a discharge rate as required in Table 2 for at least five minutes. For ships constructed on or after 1st January 2020, it is proposed that the principal agent shall be suitable for use with salt water and conform to performance standards not inferior to those acceptable to the IMO12;

Category	Helicopter overall length	Discharge rate foam solution (l/min)
H1	up to but not including 15m	250
H2	from 15m up to but not including 24m	500
H3	from 24m up to but not including 35m	800

Table 2 - Foam Discharge Rate

- at least two nozzles of an approved dual-purpose type (jet/spray) and hoses sufficient to reach any part of the helideck;
- in lieu of the requirements of points 3 and 4 above, on ships constructed on or after 1 January 2020 having a helideck, foam firefighting appliances which comply with the provisions of the IMO Fire Safety Systems Code, Chapter 17.

6 FUTURE CONSIDERATIONS

With the increase in size of yachts, and the arrival of greater demands from owners for bigger and more complex operations, the demand for multiple helidecks is now arising within the industry. Concurrent flying operations from the same vessel throws up numerous hazards that need to be carefully addressed. One of the key areas that will need to be addressed is sufficient safety equipment. To use two helidecks at the same time requires twice the amount of safety equipment. This includes First Aid Fire-Fighting Appliances (FAFFAs), rescue equipment, Fire-fighting apparatus and Fire suits. For example, consideration would need to be given to ensure that the vessel would be able to cope with the provision of firefighting medium to both Helideck concurrently. This would clearly place a high demand on High-Pressure Salt-Water Pumps etc.

Furthermore, additional problems could arise if only one of a number of helidecks on a yacht was fully certified or commercially certified, and another was not. Features within the design may have to be considered to ensure the difference was clear within the operations of the decks. Visibility of the decks from a single point, and features such as radars may also be required to ensure that crew can control the movements of more than one helicopter in the vicinity of the yacht.

7 THE CERTIFICATION PROCESS

During the design process a Certification team can work with the Design team, advising on what is certifiable. The main features of the Design Certification Assessment are to look at the Helideck itself and to evaluate layout, obstructions, deck friction levels, to identify the key flight safety features including fire hydrants and safety equipment, and to assess all equipment, procedures and documentation associated with Helicopter operations onboard. Once the criteria for the inspection have been satisfied then a Helicopter Landing Area Inspection Report (HLAIR) and a Helicopter Landing Area Certificate (HLAC) are generated and a thorough brief given to the Design team.

A HLAC covering the design activities will remain valid for as long as the design remains unchanged.

8 LESSONS IDENTIFIED

As an experienced MCA AIB, Safeguard Helideck Certification has noted several important lessons, identified from current certification activities:

- The assessment and / or certification of helidecks as part of the design process significantly reduces the risk of costly rework of the built vessel to meet requirements post build.
- For the built platform, certification is often required at very short notice due to changes of use, transport arrangements or charter requirements. Certification under these circumstances is considerably quicker, and more likely to be successful when an assessment and certification of the design has already been conducted.
- In specifying the design, the Superyacht designers may propose deck materials and features that would prevent the helideck from being certified. These may include surfaces without sufficient friction, or over height obstructions within the Limited Obstacle Sector (LOS). In such cases, if considered appropriate, materials and features may be discussed with designers to confirm that they would form a suitable basis for a certifiable helideck.

9 CONCLUSION

It is intended that this paper has illustrated some of the key aspects that should be considered by designers when incorporating a helideck into the design of a Superyacht. Consideration of these key requirements in the early stages of design will ensure customer wishes and technical requirements can be met, without unduly constraining an innovative design process. The inclusion of these key aspects within the design will contribute significantly to demonstrating that the helideck is "safe by design", regardless of whether the helideck is to be subject to full commercial certification. This will ensure the helideck is able to support safe helicopter operations.

10 REFERENCES

1. Report on air accident in Port of Bergen, Hordaland county, Norway, 10 May 2017 involving Airbus Helicopters AS 350 B3, G-HKCN.
2. International Civil Aviation Organization (ICAO) Annex 14 Volume 2.
3. Civil Aviation Publication (CAP) 437 Standards for Offshore Helicopter Landing Areas.
4. Red Ensign Group Yacht Code, Part A, Chapter 25, Annex H January 2019

11 AUTHORS BIOGRAPHY

[Chris Wood] is a Lead Helideck Certification Officer. Chris has a wealth of experience in maritime aviation having spent 15 years in the Royal Navy as Aircrew and as a Flight Deck Aircraft Handler. He has worked on numerous multi-spot flight decks including HMS Ocean, RFA Argus, HNLMS Rotterdam and HMS Illustrious. Chris retrained as Aircrew where he was part of a four-man crew flying the Merlin EH101 Mk1. Over his 15 years in the military he has worked with various types of helicopter from around the world. Chris is also a qualified Dangerous Goods by Air Instructor.

[Matt Weetch] is a Lead Helideck Certification Officer. Matt has 28 years of maritime aviation experience, 25 years in the Royal Navy and 3 years working within Industry. He was previously employed within the Military Aviation Authority, working on behalf of the Defence Naval Authority Group (Aviation). As a Maritime Aviation Professional Officer, he was responsible for delivering assurance to the Permanent Under Secretary for Defence that maritime platforms (Warships and Support Vessels) were safe to conduct complex aviation activities throughout their maintenance lifecycles.

[Steve Beech] is a Director of Safeguard Engineering Limited, operators of Safeguard Helideck Certification. Steve is a Chartered Engineer with over 25 years' experience in Systems Safety Consultancy, advising customers in the Rail, Nuclear, Commercial Marine, and Defence (Maritime, Land, and Explosives) industries. After careers with Atkins (Defence) and Frazer-Nash Consultancy, Steve set up Safeguard Engineering Limited to provide simple and cost-effective safety solutions to all aspects of the defence and commercial marine sectors.

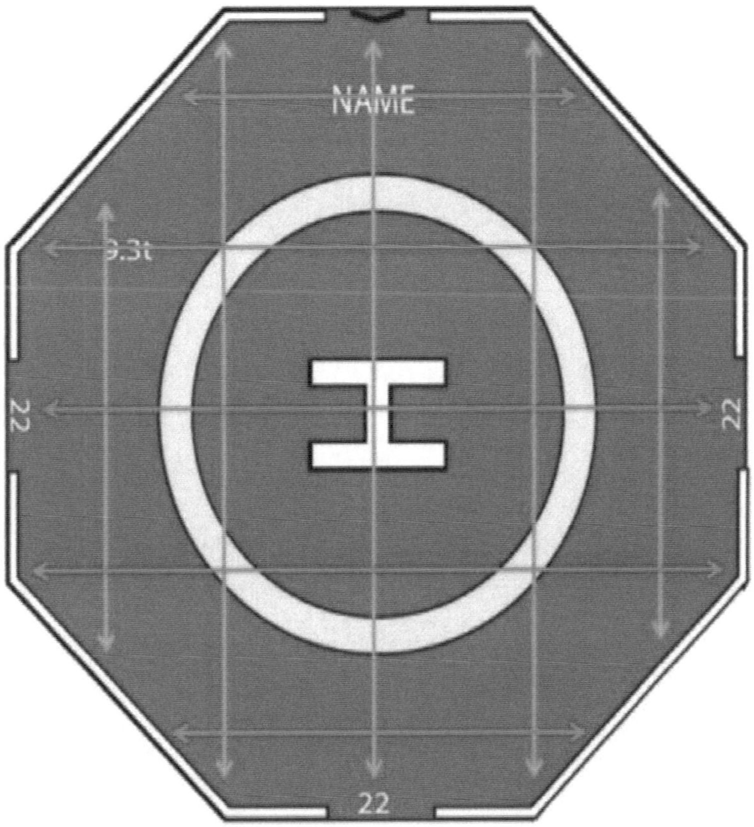

Figure 1 - Helideck Friction Survey Procedure

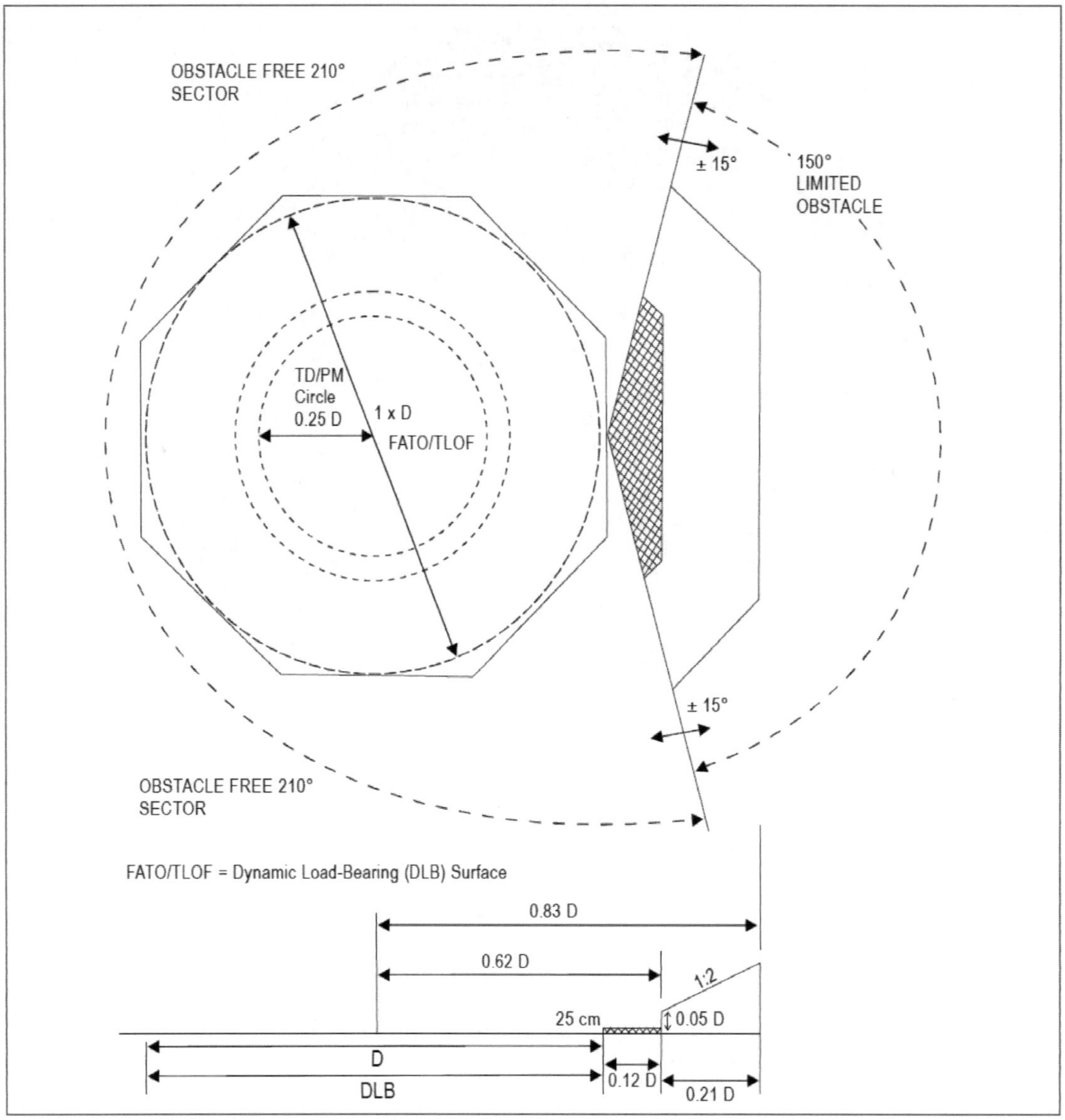

Figure 2 - Helideck Obstacle Limitation Sectors and Surfaces for a FATO and coincidental TLOF of 1D and Larger

The Royal Institution of Naval Architects
8-9 Northumberland Street
London, WC2N 5DA
United Kingdom

ISBN 978-1-5108-9538-6

CURRAN ASSOCIATES INC.
proceedings.com

First Workshop on Understanding Implicit and Underspecified Language (UNIMPLICIT 2021)

Held online due to COVID-19

Bangkok, Thailand
5 – 6 August 2021